CELL BIOLOGY RESEARCH PROGRESS

CELL PROLIFERATION

PROCESSES, REGULATION AND DISORDERS

CELL BIOLOGY RESEARCH PROGRESS

CELL BIOLOGY RESEARCH PROGRESS

CELL PROLIFERATION

PROCESSES, REGULATION AND DISORDERS

CHANGHONG ZHANG
AND
XIANGQIONG ZENG
EDITORS

NOVA BIOMEDICAL

New York

For permission to use material from this book please contact us:
Telephone 631-231-7269; Fax 631-231-8175
Web Site: http://www.novapublishers.com

Library of Congress Cataloging-in-Publication Data

ISBN: 978-1-62417-352-3

Published by Nova Science Publishers, Inc. † New York

Contents

Preface

This book was written by internationally recognized researchers and scientists from many years of their combined research and instruction work in biology, bioengineering, mathematics and pharmaceutical fields. This book is an essential practical guide to an important and expanding area about cell proliferation. A detailed practical description of cell proliferation mechanism, detection methods, and related biomaterial and drug development is provided in this book. Proliferation of plant cells is also addressed in Chapter 5 to help readers understand cell proliferation in a broad arena. I sincerely hope that this comprehensive book on cell proliferation helps scientists in basic and clinical research in understanding this important biological activity, and helps them utilize this knowledge to help improve the life quality of humans.

I am very grateful to Dr. Xiangqiong Zeng, Professor at University of Twente, also serving as the Associate Editor of this book, for her dedicated work during the long process of book editing. I also sincerely acknowledge Dr. Zeng and my family's encouragement for this work.

In: Cell Proliferation
Editors: C. Zhang and X. Zeng
ISBN: 978-1-62417-352-3

Chapter I

Glutamate and its Receptors in Controlling Proliferation of Oligodendrocyte Progenitor Cells

Maria Kukley[*]
Group of Neuron-Glia Interactions Werner Reichardt Centre for Integrative Neuroscience, University of Tübingen, Germany

Abstract

Oligodendrocyte progenitor cells, OPCs, (also known as NG2-glia, polydendrocytes or synantocytes) are widespread in the grey and white matter areas of the central nervous system. In every region of the brain, OPCs are found as proliferative cells, and both in development and in the adult at least 50% of them are actively cycling. Signals responsible for regulation of OPC proliferation are currently a topic of intense investigation. Studies in culture suggested that different molecules, including the neurotransmitters glutamate, gamma-aminobutyric acid (GABA) and acetylcholine, as well as growth factors, may contribute to regulation of OPC proliferation. Recent evidence *in vivo* demonstrated that neuronal activity can modulate proliferation and differentiation of OPCs. This chapter summarises facts indicating that the neurotransmitter glutamate may be one of the molecules involved in regulation of OPC proliferation. It also puts forward a hypothesis that glutamate released at neuron-glia synapses may mediate effects of neuronal activity on proliferation of OPCs *in vivo*.

Keywords: Oligodendrocyte progenitor cells, proliferation, glutamate, synapses, cell cycle, neuronal activity

[*] Correspondence should be addressed to: Maria Kukley. Werner Reichardt Centre for Integrative Neuroscience (CIN), University of Tübingen, Otfried-Müller-Strasse 25, 72076 Tübingen, Germany; Tel: 0049 7071 29 89180; Fax: 0049 7071 29 25006; E-mail: Maria.Kukley@cin.uni-tuebingen.de.

1. Introduction

Oligodendrocyte progenitor cells, OPCs, (also known as NG2-glia, polydendrocytes or synantocytes) are widespread in the grey and white matter areas of the central nervous system (CNS). They comprise 8–9% of the total cell population in adult white matter, and 2–3% of total cells in adult grey matter [1]. A major function of these cells is to generate oligodendrocytes [2, 3, 4, 5, 6, 7, 8, 9], glial cells which make myelin sheaths around CNS axons. Myelination of axons helps to significantly increase the speed by which action potentials are conducted and to ensure the precise timing of neuronal activity. But OPCs are interesting not only as a major source of oligodendrocytes. They are intriguing because, while being progenitor cells which are able to divide and migrate, OPCs also show features normally suggestive of differentiated cells that have essential physiological roles. Thus, unlike precursor stem cells, (just as OPCs is often short for oligodendrocyte precursor cells) OPCs possess a complex multipolar tree of fine processes [9, 10, 11, 12, 13, 14, 15], which may ramify through the neuropil for distances up to 160–200 μm [16]. Furthermore, electrophysiological studies indicate that OPCs express a complex set of voltage-gated channels including tetrodotoxin (TTX) - sensitive sodium channels and several types of potassium channels [9, 10, 11, 14, 15, 17, 18]. Furthermore, OPCs in grey and white matter areas of the brain express α-amino-3-hydroxyl-5-methyl-4-isoxazole-propionate (AMPA) /kainate and/or gamma-aminobutyric acid (GABA) receptors and receive glutamatergic and/or GABAergic synaptic input from neurons [10, 12, 19, 20, 21, 22].

Remarkably, in every region of the brain, OPCs are found as proliferative cells, and they represent the largest pool of proliferating cells in the adult CNS [1, 23, 24, 25, 26]. Different studies indicate that both in the developing and in the adult brain 50-100% of OPCs are actively cycling at any given time point [7, 14, 27], but the cell cycle time is much slower in the adult compared to the immature brain [27, 28]. Signals responsible for regulation of OPC proliferation are currently a topic of intense investigation. Studies in culture proposed that different molecules, including the neurotransmitters glutamate, GABA and acetylcholine, as well as growth factors, are involved in controlling proliferation of OPCs [29, 30, 31, 32, 33, 34, 35, 36, 37, 38]. Studies *in vivo* suggested the influence of axons on the proliferation of OPCs and indicated that neuronal activity can modulate proliferation and differentiation of these cells [39, 40, 41, 42, 43, 44, 45]. Yet, the molecular mechanisms mediating the effects of neuronal activity on OPC proliferation *in vivo* remain un-known.

This chapter summarises research related to glutamate-mediated signalling between neurons and OPCs, and emphasizes findings related to a possible role of glutamate and its receptors for proliferation of OPCs in culture and *in vivo*.

2. OPCs Represent the Largest Pool of Proliferating Cells in the Brain

During the last decade, several groups performed quantitative examination of OPC proliferation in the intact brain and spinal cord of living animals using 5-bromo-2'-deoxyuridine (BrdU) as a proliferation marker. BrdU is a synthetic nucleotide analogue, which incorporates into newly synthesized DNA of replicating cells (during the S phase of the cell cycle), substituting for thymidine during DNA replication. In these studies BrdU was given to the animals via intraperitoneal injection or supplied in the drinking water *ad libitum*, the animals were sacrificed at different time points and the number of OPCs incorporating BrdU was counted in fixed slices [1, 2, 46, 47, 48, 49, 50, 51]. BrdU was detected with anti-BrdU antibodies while OPCs were labeled with one of their specific markers, e.g. NG2 chondroitin sulfate proteoglycan or platelet-derived growth factor receptor alpha (PDGF-Rα). These experiments revealed that OPCs account for approximately 70% of BrdU positive cells in the adult cerebral cortex, hippocampus, corpus callosum and spinal cord after a short BrdU pulse [1, 46, 47, 48]. Other studies using BrdU labeling, immunohistochemistry and/or transgenic mice where NG2-positive or Olig2-positive cells are labeled suggest that virtually all cells (≥ 90%) incorporating BrdU in the brain parenchyma are OPCs [2, 49, 50, 51]. OPCs double positive for NG2 and Olig2 also represent the major cycling population of the healthy adult human brain [52].

In order to estimate the proportion of BrdU-positive OPCs within the total population of BrdU-labeled cells, single or few BrdU injections (or BrdU supplied in a drinking water for a few days) are sufficient. However, from these experiments it is difficult to infer the proliferative status of OPCs, as their growth fraction and cell cycle time cannot be determined. In order to estimate the growth fraction of OPCs, a recent study performed double staining for NG2 and the proliferating cell nuclear antigen (PCNA), which is only detected in cycling but not in resting cells [14].

At P9 and P11, the growth fraction of OPCs (double positive for NG2 and PCNA) in mouse hippocampus appeared to be as large as 48% and 49% (93/194 and 52/106 of NG2 cells), respectively [14]. These growth fraction measurements are consistent with a follow-up report indicating that the growth fraction (defined by cumulative BrdU labelling) of OPCs in corpus callosum and cerebral cortex of 6-days-old mouse is ~ 55% [27].

As an animal grows into adulthood, the density of OPCs decreases [53, 54], and the absolute number of cycling OPCs declines [54, 55]. However, although the actual number of proliferating OPCs in the adult brain is largely below that of the first postnatal weeks, the fraction of actively cycling OPCs changes only slightly with age, being ~46% in corpus callosum and ~39% in cerebral cortex of 2-18 months old mice [27].

Thus approximately half of all OPCs are constantly dividing, independently of the brain area and the age of the animals. Remarkably, recent findings indicate that *all* OPCs in the mature brain, regardless of the region, have the ability to divide [7]. Therefore, it is possible that the total number of cycling OPCs in the brain is even higher than suggested previously [14, 27].

Notably, proliferation parameters of OPCs differ significantly between the neonatal and the adult brain. For instance, cumulative BrdU labeling performed in mice of various ages revealed striking discrepancy in cell cycle time of OPCs [3, 14, 27]. During the first two postnatal weeks the cell cycle time of OPCs in the hippocampus, cerebral cortex and corpus callosum is about 2-3 days, meaning that each cycling OPC divides ~ every 48-72 hours [14, 27].

As the animal matures, the cell cycle time increases steadily, being > 100 days at P54 [27]. Interestingly, the rate of oligodendrocyte production from OPCs declines in parallel with the increase of the OPC cell cycle duration. In the corpus callosum, for example, the cell cycle slows down ~ 10-fold between P45 and P240 and the rate of oligodendrocyte production slows ~ 20-fold at the same period [27].

3. Glutamate Receptors on Oligodendroglial Progenitors and their Role during Cell Proliferation in Culture

The amino acid glutamate is a major excitatory neurotransmitter which accounts for most of the fast synaptic transmission that occurs in the mammalian CNS. During brain development, glutamate mediates the effects of electrical activity on proliferation, migration and differentiation of neurons [56]. Glutamate exerts its action via different types of ionotropic and metabotropic receptors present on neurons. Ionotropic receptors are membrane-bound protein complexes typically composed of several individual proteins that combine to form an ion channel through the membrane [57]. Ionotropic glutamate receptors are classified into 3 sub-types: N-methyl-D-aspartate (NMDA), 2-amino-3-(5-methyl-3-oxo-1,2-oxazol-4-yl) propanoic acid (AMPA), and kainate receptors. Metabotropic glutamate receptors are composed of a single polypeptide and exert their effects not through the direct opening of an ion channel but rather by binding to and activating G-proteins [57]. The metabotropic receptors are coupled to inositol phospholipid metabolism, cyclic adenosine monophosphate (cAMP) formation and mobilization of intracellular calcium ($[Ca^{2+}]_i$) via G-proteins [58].

The first demonstration of glutamate-activated channels in glia goes back to 1989, when multiple conductance channels activated by L-glutamate, quisqualate or kainate were reported in type-2 astrocytes in culture [59]. At that time, it was suggested that glial cells possess glutamate receptors which could be activated by glutamate released from axons and may, therefore, be important for formation and maintenance of the nodes of Ranvier [59]. Follow-up studies revealed that glutamate, kainate or quisqualate, agonists at ionotropic AMPA/kainate receptors, elicit inward currents in oligodendrocyte-type-2 astrocyte (O-2A) progenitor cells (presumably OPCs) acutely isolated from optic nerve [60], in cultured cerebellar and cortical O2-A progenitors [61, 62], and in glial precursor cells in the corpus callosum *in situ* [63].

In the search for a possible function of these glial glutamate receptors, it has been suggested that they may be involved in the regulation of proliferation of oligodendroglial progenitor cells. Several studies have addressed this question using dissociated cell cultures and organotypic slice cultures [29, 30, 31, 32, 64]. It turned out that agonists of glutamate receptors, glutamate and kainate, cause a concentration dependent decrease in [^{3}H]thymidine incorporation in brain oligodendrocyte progenitors in dissociated cultures, with an IC_{50} of 100 μM and 30 μM, respectively [64]. Although the mechanism of this effect has not been studied in detail, it has been suggested that protein kinase C (PKC) participates in the cascade of events activated by glutamate, which ultimately leads to regulation of gene expression [64]. Independent findings from another group also demonstrated that activation of glutamate receptors by glutamate (200 μM), kainate (200 μM) or AMPA (200 μM) inhibits proliferation of cortical O-2A progenitors (presumably OPCs) in dissociated cultures [29]. An antagonist of ionotropic non-NMDA type glutamate receptors, 6,7-dinitroquinoxaline-2,3-dione (DNQX, 30 μM), prevented the antiproliferative effects of glutamate receptor agonists [29]. These results suggested that ionotropic non-NMDA glutamate receptors mediate the antiproliferative effect of glutamate on cultured oligodendrocyte progenitors. Furthermore, glutamate receptor agonists also influenced OPC proliferation through selective activation of AMPA receptors in a cytoarchitecturally intact system – cerebellar tissue slices [31]. Gallo and colleagues suggested that the triggering mechanism by which activation of glutamate receptors inhibits proliferation of OPCs, both in cultured cells and in cultured slices, is by raising the intracellular Na^+ concentration, and thus inhibiting voltage-dependent K^+ channels [29, 30, 31]. Consistent with a role for K^+ channels in regulating proliferation of OPCs, the over-expression of the Kv1.3 or Kv1.4 subunits of the voltage-gated K^+ potassium channel enhanced oligodendrocyte lineage cell proliferation [65], while application of tetraethylammonium, which blocks K^+ channels, inhibited proliferation [29, 31]. Importantly, similarly to activation of glutamate receptors, agents that directly increase intracellular Na^+ or depolarize the cell membrane caused marked antiproliferative effects on cultured O-2A progenitor cells through the reduction of voltage-dependent outward K^+ currents [30, 31]. For example, elevation of intracellular Na^+ with the alkaloid veratridine or increase in extracellular concentration of K^+ ions causes a reduction of K^+ currents in O-2A cells and markedly inhibits their proliferation even in the presence of mitogenic factors PDGF and basic fibroblast growth factor (bFGF) [30, 31].

Later experiments in culture provided evidence that activation of glutamate receptors by 100 μM kainate also inhibits proliferation of O4-positive oligodendrocyte progenitors derived from epidermal growth factor (EGF)-expanded rat striatal neural stem cells [32]. In contrast to earlier findings [29, 30, 31], however, the decrease in BrdU incorporation in striatal O4-positive cells promoted by 100 μM kainate was mediated by kainate receptors, and not by AMPA receptors, because the AMPA/kainate receptor antagonist 6-cyano-7-nitroquinoxaline-2,3-dione (CNQX) completely abolished this effect, but an antagoist of only the AMPA receptor, 4-4(4-aminophenyl)-1,2-dihydro-1-methyl-2-propylcarbamoyl-6,7-methylene dioxy phthalazine (SYM 2206), did not prevent it [32]. The authors proposed that 100 μM kainate might inhibit the proliferation of O4-positive oligodendroglial progenitors to induce their apoptotic cell death by increased Ca^{2+} influx and Ca^{2+} accumulation in mitochondria. Interestingly, application of kainate in lower concentrations (1 and 10 μM) had the opposite effect: BrdU incorporation in O4+ striatal progenitors was significantly increased. This effect was abolished in the presence of SYM 2206 indicating that it is mediated by AMPA receptors [32]. The mechanism of this mitogenic effect involved the increased phosphorylation of the cAMP response element binding protein (CREB) by protein kinase A and C [32]. These findings are in line with previous reports which showed that CREB phosphorylation by protein kinase A and C is associated with proliferation of oligodendrocyte progenitors [66, 67, 68], and that a significant stimulation of CREB phosphorylation occurs upon incubation of cultured oligodendroglial progenitors with the non-NMDA glutamate receptor agonists kainate or glutamate [66, 69].

Taken together, data obtained in cell culture suggest two possible mechanisms by which glutamate may control proliferation of OPCs. Glutamate, acting via AMPA/kainate receptors on OPCs may inhibit proliferation by a mechanism which involves changes in the cell membrane potential and/or an increase in intracellular sodium concentration, leading to inhibition of voltage-dependent K^+ channels. On the other hand, glutamate binding to

AMPA/kainate receptors on OPCs may also stimulate their proliferation by triggering activation of protein kinases and a subsequent cascade of events, leading to increased CREB phosphorylation and modulation of gene expression, e.g. genes involved in cell cycle progression. At present, it remains unclear which factors determine whether glutamate stimulates or inhibits proliferation of OPCs in culture.

Notably glutamate is not the only neurotransmitter involved in regulation of OPC proliferation. Exposure of cultured oligodendroglial progenitors to baclofen, an agonist of metabotropic GABA receptors ($GABA_B$-receptors) increased their proliferation, providing evidence for a functional role of $GABA_B$ receptors in oligodendrocyte development [33]. Carbachol, a stable acetylcholine analogue, has been shown to stimulate DNA synthesis in oligodendrocyte progenitors [34, 35]. The effect of carbachol is mediated by muscarinic acetylcholine receptors (mAChRs) because it is prevented or significantly reduced by the addition of the muscarinic antagonist atropine [34, 35]. Further evidence for involvement of mAChRs comes from experiments demonstrating that treatment with muscarine induces an increase of ^{3}[H]-thymidine incorporation levels which was counteracted by atropine [36]. The M1 receptor antagonist pirenzepine, M3 receptor antagonist 4-diphenylacetoxy-N-methylpiperidine-methiodide (4-DAMP) and M4 receptor antagonist tropicamide significantly reduced muscarine-induced ^{3}[H]-thymidine incorporation, suggesting that these three receptor subtypes mediate cholinergic effects on oligodendrocyte progenitors proliferation [36]. Adenosine triphosphate (ATP) and adenosine diphosphate (ADP) have been shown to inhibit proliferation of oligodendrocyte progenitors induced by PDGF, both in purified cultures and in cerebellar tissue slices [37]. The effects of ATP and ADP on cell proliferation were prevented by the purinergic receptor Y1 (P_2Y1) receptor antagonist N6-methyl-2'-deoxyadenosine-3',5'-bisphosphate (MRS2179) [37], indicating that also metabotropic P_2 receptors are able to control development of oligodendrocyte lineage cells. Treatment with another neuroligand - adenosine (1–300 μM) and the general adenosine receptor agonist 5'-N-ethylcarboxamidoadenosine (NECA, 1–300 μM) caused a substantial, concentration-dependent decrease in proliferation of cultured oligodendrocyte progenitors [38]. Chronic treatment (48 hr) of cerebellar slices with adenosine (100 μM) also significantly decreased the percentage of lymphoblast antigen 1 (LB1)-positive or NG2-positive oligodendroglial progenitors that incorporated BrdU as compared with untreated slices [38]. Some other neuroligands, including substance P, histamine, and norepinephrine, are known to evoke Ca^{2+} signals in oligodendrocyte progenitors [70, 71], and therefore could potentially also influence proliferation of these progenitor cells, although their role has not been yet directly demonstrated.

4. Glutamatergic Synapses on Oligodendroglial Progenitors and their Possible Role during Proliferation

If glutamate plays a key role in regulating proliferation of oligodendroglial progenitors not only in culture but also under normal physiological conditions *in vivo*, then sources of glutamate to activate glutamate receptors on oligodendroglial progenitors should be constantly available in the brain. Glutamate release under physiological conditions is usually attributed to fusion of synaptic vesicles in neurons at neuronal synapses where it mediates synaptic signaling and plasticity [72]. In addition, glutamate can also be secreted by glial cells suggesting that glia may be an important point source for ambient extracellular glutamate [73, 74]. However, a powerful glutamate uptake system keeps the ambient extracellular concentration of glutamate in the brain low [75, 76]. The tonic spatially and temporally averaged concentration of extracellular glutamate in acute brain slices with intact glutamate transport is about 25-50 nM [77, 78]. Microdialysis *in vivo* yields higher estimates for extracellular glutamate that are typically in the range of 1 to 5 μM after correction for recovery [76]. Native AMPA receptors in neurons are activated by glutamate concentrations between ~100 μM and ~10,000 μM, with an EC50 (effective concentration required to induce a 50% effect) of ~700 to ~1000 μM [76]. Kainate receptors are activated at slightly lower glutamate concentrations, with EC50s of ~300 to ~800 μM [76]. Assuming that AMPA/kainate receptors on oligodendroglial progenitors have similar affinities, neither AMPA nor kainate receptors are likely to be activated by ambient extracellular glutamate (even if the higher estimates of ambient glutamate concentration are correct). Hence, a condition when AMPA/kainate receptors on oligodendrocyte progenitors are constantly exposed to a glutamate receptor agonist (as it is the case in culture studies, where glutamate receptor agonists are continuously present in the medium in concentrations high enough to activate AMPA/kainate receptors), is un-likely to occur in the healthy brain *in vivo.* Yet, a specific activation of AMPA/kainate receptors on OPCs occurs *in vivo;* the brain has accomplished this task by establishing neuron-glia synapses.

4.1. OPCs Receive Functional Synaptic Input from Neurons

Synapses between neurons and OPCs were first discovered by Bergles and colleagues [10]. The authors made whole-cell patch-clamp recordings from OPCs in the CA1 (*Cornu ammonis* 1) region of the hippocampus and measured the response of these cells to stimulation of afferent excitatory axons, Schaffer collaterals. They could record inward currents in OPCs which were inhibited by the AMPA/kainate receptor antagonists [10]. The quantal nature of these responses and their rapid kinetics indicated that they are produced by the exocytosis of vesicles filled with glutamate directly opposite AMPA receptors on oligodendroglial progenitors [10]. A similar type of fast neuron-glia synaptic signaling has now also been discovered between GABAergic interneurons and OPCs in the hippocampus [19]. Additional evidence from electron microscopy analysis revealed accumulations of small vesicles at contact sites between axons and the processes of OPCs in both the young and the adult rodent hippocampus [10, 19].

Since the first discovery of glutamatergic and GABAergic innervation of OPCs, several groups have demonstrated that these cells receive synaptic input from neurons in different grey matter areas including the cerebral cortex, hippocampus, cerebellum, brain stem and ventrobasal thalamus [11, 14, 17, 18, 19, 20, 79, 80, 81]. More surprisingly, un-myelinated axons passing through white matter (e.g. corpus callosum, optic nerve) also establish excitatory glutamatergic synapses with local oligodendroglial progenitors, and neurotransmitter release along axons in white matter is quite similar to vesicle fusion at synapses: it is reliable, it depends on action potential propagation, involves highly localized calcium microdomain signalling and is strongly calcium cooperative [12, 22].

In summary, many if not all OPCs in the developing and mature brain are engaged in synaptic signaling with neurons. At the same time, a large proportion, if not all OPCs, regardless of region, has the ability to divide. Considering the immense biochemical effort to set up a synaptic junction, the question arises whether synaptic junctions are simply discarded each time an OPC performs a cell division. If OPCs disassemble all synapses before cytokinesis and build them up *de novo* after cell division, it is puzzling how newborn OPCs can rapidly get connected to functional release sites, considering that newborn neurons need about 3 days to acquire synaptic input [82]. An alternative possibility is that OPCs largely preserve their synaptic connections with neurons during cytokinesis. This would most likely mean that these cells divide without retracting their processes, because AMPA receptors involved in synaptic signaling are located along the processes of the cell [83]. Up to now, few studies have addressed the question of whether OPCs keep their processes and maintain synapses with neurons during cell division [9, 14, 84, 85].

4.2. Morphology of OPCs during Mitosis

To investigate the morphology of proliferating OPCs *in situ*, patch-clamp recordings from individual hippocampal and cortical OPCs with a mitotic DNA configuration have been carried out in brain slices and the cells were filled with a fluorescent dye Lucifer Yellow [14]. Mitotic cells are hardly visible if observed only with differential interference contrast of transmitted light mode. Therefore, for labeling mitotic cells, individual live slices were incubated for 10 min with a membrane permeable fluorescent DNA stain Hoechst 33342 (0.1 mg/ml) before recordings. Cells were then selected according to presence of mitotic figures of their DNA using conventional fluorescence microscopy. Patch-clamp recordings were obtained from solitary cells showing a metaphase arrangement of chromosomes, and from paired cells with a mitotic DNA configuration, which were most likely to be in the telophase of mitosis [14]. To verify that the recorded cells were OPCs, post-recording staining for NG2 was carried out. Subsequent morphological analysis of Lucifer Yellow-filled NG2-positive cells revealed that metaphase and telophase OPCs possess a rich tree of branching processes (Figure 1), [14]. Similar results have been reported by Ge and colleagues who carried out morphological examination of mitotic OPCs positively labeled for NG2 in brain slices [85]. These findings suggest that OPCs maintain their complex morphology during cell division. Yet, they do not completely exclude the possibility that OPCs are able to retract and re-grow some of their processes very quickly, e.g. on the range of minutes, and at the moment of chromosome separation the cells withdraw many of their processes. Time-lapse imaging of zebrafish OPCs *in vivo* indeed showed that cells can rapidly remove filopodium-like processes, divide and then re-grow their processes again and migrate away from each other [84]. *In vivo* time-lapse imaging of mammalian OPCs has not been reported to date and it is unclear whether a similar sequence of events happens during the cell division of mammalian oligodendroglial progenitors. Ge and colleagues attempted to perform time-lapse imaging of dividing OPCs in acute mouse brain slices and observed that the soma of the mother cell was splitting into two, while all of the processes remained un-retracted [85]. This result strongly supports the idea that mammalian oligodendrocyte progenitors can retain their complex morphology during mitosis. However, the observation of Ge and colleagues is so far based only on a single example, and further experiments involving time-lapse imaging in transgenic mice where the soma and processes of OPCs are specifically labeled are required.

The ability of OPCs to preserve their complex morphology during cell division may appear unusual because we know that many cells retract processes and round up during mitosis [84, 86]. Nevertheless the presence of processes on some proliferating cells has been described previously. Video time-lapse observations suggest that in the superior cervical ganglion about 5% to 10% of neuron precursors show an unusual differentiated morphology and are able to retain long processes over 200 μm in length during mitosis [86]. Another study illustrates that in the retina of p107-single (Chx10-Cre; Rblox/lox; p107+/−; p130−/−) transgenic mice fully differentiated horizontal cells are capable of re-entering the cell cycle and keep their neurites and synapses while dividing and clonally expanding [87]. Mitotic cells displaying all the features of differentiating motor neurons, including axonal projection via the ventral root have also been described in the spinal cord of a chick [88]. It remains unclear why some cells preserve more or less complex morphology during mitosis while the others do not. For neuronal progenitors, an advantage of elaborating processes while maintaining the capacity to divide could lie in the fact that early target tissue contacts may influence neuroblast cell division [86]. In the case of a glial cell, keeping processes and synaptic contacts (see below) with neurons during mitosis could be necessary as it might give neurons the possibility to regulate proliferation of oligodendroglial progenitors and possibly influence their developmental fate.

4.3. Neuron-glial Synapses during Mitosis

To investigate synaptic responses directly in proliferating cells, patch-clamp recordings have been performed from mitotic OPCs (metaphase or telophase of mitosis) in acute hippocampal and cortical slices from neonatal (7-12 days old) mice [14]. In the presence of the $GABA_A$ receptor antagonist bicuculline (10 μM), fast rising and fast decaying AMPA/kainate-receptor mediated currents could be recorded in OPCs during metaphase and telophase of mitosis (Figure 1), [14]. These results indicate that shortly before and shortly after cytokinesis OPCs possess functional synaptic junctions with neurons. In line with these findings are the recent data of Ge and colleagues, who demonstrate that grey and white matter OPCs in metaphase and telophase of mitosis receive glutamatergic synaptic input from neurons not only in young but also in older animals (up to 20 weeks of age) [85].

Based on the described evidence it is plausible that oligodendroglial progenitors in the rodent brain can enter cell cycle and undergo cell division without losing functional glutamatergic synapses with neurons. The evidence so far is indirect because single synapses on OPCs during mitosis have not been tracked in living brain slices or *in vivo*. However, if we assume that the time necessary for the assembly and disassembly of neuron-glia synapses is the same as for neuronal synapses, then AMPA receptor-containing synapses can be eliminated as quickly as 90 min and assembled within 1 to 2 hours of initial axo-dendritic contact [89, 90, 91]. If OPCs have to disassemble *all* synapses before cell division and reform *all* of them from scratch after cytokinesis, there should be a period of time equal to at least 1 hour when synaptic currents are absent in OPCs. However, synaptic currents could be recorded in OPCs 30 minutes before and after chromosome segregation [14, 85].

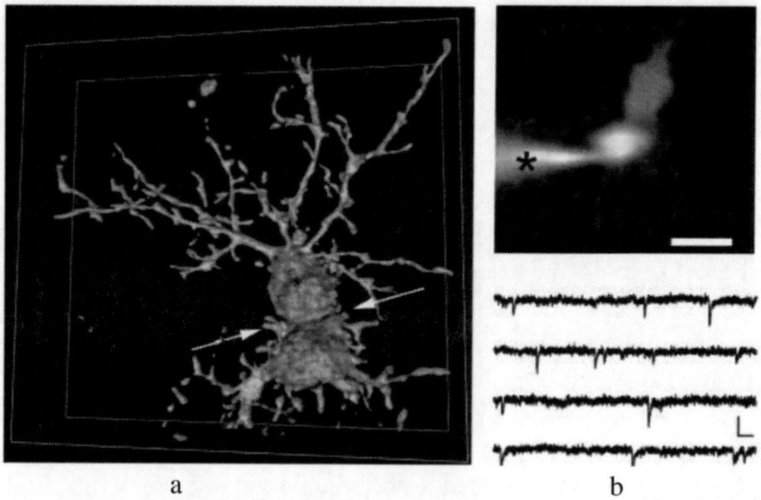

Figure 1. OPCs in telophase of mitosis show a complex tree of processes and receive synaptic input from neurons. (A) 3D semitransparent reconstruction of a grey matter OPC. Note mitotic DNA bulging out of the somatic cell membrane. Arrows indicate prominent division furrow. Dimensions of the cube shown are 43μm x43μm x12 μm. (B) Top: mitotic OPC filled with Cy5-conjugated dextran (10 kDa) during patch-clamp recordings. Dextran (10 kDa) does not spread through gap junctions, so filling of two cells indicates that the division process is not completed and cells still share the cytoplasm. Asterisk indicates the patch-pipette. Scale bar: 10 μm. Bottom: glutamatergic synaptic currents recorded from the cell shown at the top (holding potential is -80 mV), in the presence of $GABA_A$ receptor antagonist bicuculline (10 μM). Scale bar: 20 ms, 10 pA. Modified with permission from [14].

Therefore, it is likely that OPCs maintain synaptic connections with neurons during cytokinesis, and in this way newly generated OPCs can inherit synapses from their parent cell in the postnatal rodent brain. What could be the benefit of inheriting a synapse? Keeping synaptic contacts with dividing OPCs may give neurons the opportunity to directly or indirectly regulate proliferation of OPCs, e.g. by influencing the expression of potassium channels during the cell cycle and/or by controlling other intracellular signaling cascades. On the other hand, inheritance of synapses may allow for the direct transfer of environmental interactions to clonal descendants of OPCs. This might be important for effective colonization and perhaps future myelination of the developing brain.

4.4. Are Neuron-glia Synapses Involved in Regulation of OPC Proliferation?

Fast vesicular release of glutamate at neuron-glia synapses followed by binding of glutamate to AMPA receptors on OPCs is probably the most usual way of activating AMPA receptors on OPCs in the healthy brain. In culture, pharmacological activation of neurotransmitter receptors on OPCs influences their proliferation [29, 30, 31, 32]. If the AMPA receptor mediated effect of glutamate is involved in the regulation of cell proliferation also *in vivo,* then synaptic junctions between neurons and OPCs are expected to be the morphological and functional correlates of this regulation. Yet, the functional significance of neuron-glia synapses and their role in cell proliferation remains unknown. The evidence regarding involvement of glutamate and its receptors in controlling oligodendrocyte progenitor proliferation obtained in culture cannot be directly transferred to the *in vivo* (or *in situ*) situation. A major reason for this is that the modes of AMPA/kainate receptor activation on OPCs *in vivo* and in culture are most likely quite distinct. *In vivo,* AMPA-receptor mediated signaling between neurons and OPCs depends on the strength and time-course of synaptic neuron-glia responses. These parameters are in turn determined by factors that fall into three categories: those affecting the time course of transmitter release at neuron-OPC synapses, those affecting the lifetime of transmitter in the synaptic cleft (i.e., diffusion and uptake), and the number and properties of postsynaptic receptors on OPCs (i.e., gating kinetics, desensitization, deactivation and transmitter affinity) [92]. Vesicular glutamate release at neuronal synapses is fast, but glutamate lifetime in the cleft is short because, after being released, glutamate rapidly diffuses away from its release site and/or is removed by glutamate transporters. Assuming neuron-glia synapses work similarly to neuronal synapses, glial AMPA receptors *in vivo* are exposed to very short (millisecond range) pulses of synaptically released glutamate. In culture, glutamate receptor agonists (e.g. kainate, quisqualate, AMPA) are constantly present in the culture medium; they are un-likely to be taken up by the transporters and can not diffuse away from the receptors because there is no concentration gradient in the medium. Therefore, in culture, AMPA/kainate receptors on OPCs are continuously exposed to the agonist present in the medium in a defined concentration [29, 30, 31, 32]. Glutamate and also other glutamate receptor agonists such as kainate and quisqualate trigger both activation and desensitization of AMPA/kainate receptors [93, and citations therein]. Native AMPA and kainate receptors have desensitization EC50 values for glutamate of ~4 μM and ~3 to ~13 μM, respectively [76]. Activation is faster than desensitization, and thus, typically precedes desensitization when glutamate concentrations rise quickly [76], as it is the case during fast neurotransmitter release at synapses. If the agonist is kainate (as it is often the case in culture studies) many AMPA receptors may never activate because they desensitize without opening [93, and citations therein]. Considering the prolonged (hours) exposure to glutamate receptor agonists in culture, the mode of AMPA/kainate receptor activation on OPCs in culture is unclear. The receptors may get transiently activated and then desensitize, or they may get desensitized without opening, or they may be activated continuously for a prolonged period of time. This is likely to depend on the agonist, the subunit composition of AMPA receptors, the trans-membrane AMPA receptor regulatory proteins, and perhaps other factors [93].

Beside this, there are other important aspects complicating the interpretation of the results for the role of glutamate in cell proliferation obtained in culture. In particular, different factors present in culture medium (e.g. growth factors) may influence expression and function of glutamate receptors on oligodendroglial progenitors and affect synaptic modulation in cultured brain slices. In cultured brain slices, neuronal glutamate receptors are likely to be affected as well when an AMPA/kainate receptor agonist is present in the medium.

Another important issue to consider is that in order for glutamatergic inputs to regulate OPC proliferation *in vivo*, glial excitatory postsynaptic currents arising in different cellular compartments (e.g. cell body versus distant cellular processes) must be integrated in the cell soma, where patterns of activity are likely to depend not only on the number of synapses being activated, but also on their anatomical position on the OPC [13]. The mechanism of integration of excitatory postsynaptic currents in the OPC is likely to depend on spatio-temporal summation, and the summation is stronger if the excitatory postsynaptic currents in the OPC are more synchronized [13]. At present, it remains unknown how excitatory postsynaptic currents are integrated in OPCs, and whether synchronized activity actually inhibits or promotes cell proliferation. Furthermore, the influence of glutamatergic inputs on cell proliferation *in vivo* must be seen in concert with other signals that may either promote or inhibit cell division [13], and these signals are not yet elucidated.

Thus, it remains to be discovered whether and how glutamatergic synaptic signaling between neurons and OPCs regulates proliferation of these glial cells *in vivo*. The recent generation of several transgenic mouse lines, in which Cre-recombinase is expressed and inducible in oligodendroglial cells, e.g. NG2-Cre, NG2-CreERT2, PDGFRα-CreERT2, CNP-Cre and PLP-Cre lines, makes it possible to specifically delete selected genes in these cells at different timepoints [3, 4, 7, 18, 94, 95, 96]. Ablation or modification of AMPA receptors, or critical postsynaptic anchoring proteins such as e.g. PSD-95 (postsynaptic density 95), specifically in oligodendrocyte progenitors may help elucidating the functional role of glutamatergic neuron-glia synapses for cell proliferation.

5. Proliferation of Oligodendroglial Progenitors *In Vivo* and its Modulation by Electrical Activity in Axons

While research on the role of glutamate for proliferation of OPCs is still in its infancy, it has already brought interesting discoveries, and two important pieces of information are now emerging. First, glutamate may be involved in controlling proliferation of OPCs and this effect is mediated through AMPA/kainate receptors. Second, glutamatergic synapses between neurons and OPCs may represent cell-cell interaction points where glutamate-mediated regulation of cell proliferation is implemented. An essential question to ask is when and why regulation of glial cell proliferation by glutamate and glutamatergic neuron-glia signaling would be required *in vivo*. Depending on patterns of neuronal activity, glutamatergic signaling at neuron-glia synapses will most likely vary. Neurons may use neuron-glia synapses to tune the development (e.g. proliferation, migration, differentiation) and/or function of OPCs based on neuronal activity. Although still speculative, this hypothesis is worth considering. Oligodendroglial cells myelinate axons ensuring enhanced nerve conduction. During development oligodendrocytes (or possibly even their precursors) have to recognize axons, distinguish them from other tubular structures in the neuropil, and to establish contacts with selected axons, which they enwrap with myelin sheaths. Hence it would make logical sense for axonal activity to play an important role in controlling the development of oligodendrocyte lineage cells [97]. In rodents, myelination of CNS axon tracts starts postnatally and is largely completed during the first 3-4 weeks of life. In order to achieve rapid myelination of axons it is necessary to rapidly generate large numbers of progenitors, and oligodendrocytes, within a relatively short period of time. How the brain accomplishes this task remains unknown, and it has been suggested that axons regulate proliferation of oligodendroglial progenitors directly or indirectly providing mitotic signals. Glutamatergic neuron-glia synapses could be structural and functional correlates of neuronal regulation of cell proliferation.

5.1. Neuronal Activity Controls Proliferation of Oligodendroglial Progenitors *In Vivo*

In 1993, Barres and Raff showed that axons control the rate of proliferation of developing OPCs in the optic nerve *in vivo* [41]. When developing optic nerves are transected, the number of mitotic OPCs falls by 90% in 4 days. This same percentage reduction was obtained regardless of whether the total number of mitotic OPCs per longitudinal section, the total number of BrdU-positive OPCs per entire optic nerve, or the number of all OPCs per optic nerve was measured [97]. Intraocular injection of TTX, which silences the electrical activity of retinal ganglion cells and their axons, decreases the number of OPCs by 80% [41].

Further evidence which revealed a powerful role of neuronal activity in controlling oligodendrocyte progenitor proliferation *in vivo* was obtained using experimental paradigms where neuronal activity is modified by mean of direct electrical stimulation of fiber tracts [42], experimental electroconvulsive seizures [39, 40] or a specific behavioral performance [43, 44, 45].

Li and colleagues stimulated unilateral corticospinal axons of the adult rat and investigated proliferation and differentiation of OPCs in dorsal corticospinal tract using double -labeling with a mitotic indicator BrdU and specific cellular markers. Electrical stimulation of the pyramidal tract increased proliferation and differentiation of OPCs contacting or in close proximity to neighboring axons within the pyramidal tract in the spinal cord [42]. Remarkably, the effect of electrical activity was specific for the oligodendrocyte lineage cells, as no proliferation of astrocytes or endothelial cells was observed.

Increased proliferation of oligodendroglial progenitors (positive for NG2) in several brain regions of the rat (amygdala, hippocampus) has been observed in response to electro-convulsive seizures [39, 40], a paradigm clinically used for the treatment of severe depression. Rats were subjected to five treatments with electroconvulsive seizures (0.5 sec @ 50 Hz), injected with BrdU, and the brain sections were double-stained for BrdU and the cell-type markers. In response to electroconvulsive seizures, the number of BrdU-positive/NG2-positive cells in the central, lateral, and basal amygdala nuclei increased by 84%, 129%, and 132%, respectively [40]. BrdU-labeled NG2-expressing oligodendroglial progenitors were still present 3 weeks later, and a small proportion of the proliferating cells had differentiated into mature oligodendrocytes [40]. The proliferation of NG2-positive OPCs was also enhanced in the hippocampus after electroconvulsive seizures: after five seizure trials the number of proliferating OPCs increased with 248%, 424% and 528% compared to control animals in the molecular layer, hilus, and granular layer of the dentate gyrus, respectively [39].

Proliferation of OPCs also increased in response to behavioral stimuli promoting activity, such as voluntary wheel running, and environmental enrichment [43, 44, 45]. Thus, mice living in a large cage (80×80 cm^2) with nesting material, mesh wire ladders, and plastic tubes which can be rearranged, showed a significant increase of NG2/BrdU double-positive cells in the amygdala [44]. The number of proliferating oligodendroglial progenitors (double-positive for BrdU and NG2, or PDGFRα, or Olig2) was also higher in the amygdala and hippocampus of mice which were housed under standard conditions but were given *ad libitum* access to a running wheel, a voluntary exercise that leads to an increase in neuronal activity, compared to animals that did not have the possibility to run in a wheel [44, 45].

Another interesting observation made by several groups is that chronic stress affects proliferation of OPCs [98, 99]. A reduction of NG2-positive oligodendroglial progenitors incorporating BrdU has been observed in several cortical areas (prefrontal, piriform, cingulated and motor cortex), as well as in the hippocampus, in animals subjected to different chronic stress paradigms [98, 99]. Stressors included: cold (4°C), swim stress, rotation, isolation overnight, food/water deprivation, light overnight [99], as well as chronic social stress based on the resident–intruder paradigm [98]. Chronic unpredictable stress also reduced the number of mature oligodendrocytes [99], indicating that stress can affect myelination in the brain. Remarkably, chronic stress-mediated dysfunction is known to be accompanied by dysregulation of the pattern of synapsin and synaptophysin expression, decrease in the complexity of neurite arborisation, loss of synapses, and reduction in neurotransmitter release [100].

Thus, evidence is starting to emerge that an increased proliferation of oligodendroglial progenitors *in vivo* accompanies increased neuronal activity, while reduced proliferation of these cells goes along with a reduction of synaptic signaling and plasticity.

5.2. How do Electrically Active Axons Signal to Glial Cells to Divide?

Glutamate is released at synapses between neurons and OPCs and is therefore ideally suited to mediate the effects of neuronal activity on proliferation of oligodendrocyte progenitors *in vivo*. OPCs in several brain regions possess Ca^{2+}-permeable AMPA receptors [10, 22, 101], and increased activation of these receptors during enhanced neuronal activity may be followed by increased CREB phosphorylation by protein kinases [102]. The transcription factor CREB is the target for both calcium and cAMP signals, and is rapidly phosphorylated in response to depolarization or cAMP, at a site known to be important for the transcriptional activating function of this protein [103]. Increased CREB phosphorylation caused by neuronal activity may lead to changes in transcription of specific genes, in particular genes involved in cell cycle regulation [104], in OPCs. Diminished phosphorylation of CREB accompanying decreased neuronal activity would be, in turn, expected to down-regulate cell proliferation. Consistent with this hypothesis, a positive correlation exists between reduced cell proliferation [98, 99] and a defect in CREB phosphorylation [100] during chronic stress.

Apart from glutamate-mediated synaptic signaling between neurons and glia, there are likely to be other mechanisms that could explain how increased neuronal activity drives glial cell proliferation in the brain. These mechanisms can be divided into two major groups: mechanisms involving neuron-glia signaling by diffusible substances (including glutamate), and mechanisms involving direct contact between axons and OPCs. Substances that could be released from axons during activity and directly engage in neuron-glia signaling in the brain include ATP, and its breakdown product adenosine, acetylcholine, GABA, and perhaps also some other neuroligands. OPCs express receptors for those neuroligands, show calcium increases upon application of these substances in culture and in slices [70, 105, 106], and also show increased incorporation of the proliferation markers BrdU or [^{3}H]thymidine upon activation of $GABA_B$, acetycholine, P1 or P2 receptors [33, 34, 35, 36, 37, 38].

Substances released by active axons may also modulate proliferation of OPCs indirectly. For example, active axons have been suggested to release a signal that stimulates astrocytes to release PDGF which is a mitogen of OPCs [41].

Additionally, or alternatively, mechanisms involving direct contact of axons and OPCs may play an important role in regulating activity dependent proliferation of oligodendrocyte progenitors. A recent study showed the ability of an axolemma-enriched fraction (AEF) isolated from mature CNS to influence the proliferation of OPCs [107]. Addition of AEF to cultured oligodendroglial progenitors resulted in a dose- and time-dependent increase in proliferation that was partially dependent on protein kinase B and mitogen-activated protein kinase (MAPK) activation [107]. The major mitogen was identified as acidic fibroblast growth factor and accounted for 50% of the mitogenicity [107]. The remaining 50% of the mitogenicity had properties consistent with basic fibroblast growth factor, but this was not unequivocally identified [107]. As, following addition of AEF to cultured oligodendroglial progenitors, the AEF was found localized to the outer surface of their cell bodies and processes, it was suggested that the effects of AEF on proliferation and survival were mediated by direct contact between AEF and oligodendroglial progenitor cells [107].

Conclusion

This chapter attempts to gather facts indicating that neurotransmitter glutamate may be one of the molecules involved in regulation of OPC proliferation. While many ideas in this chapter are speculative, and perhaps oversimplified, they aim to inspire thinking about new experiments to reveal mechanisms regulating proliferation of OPCs *in vivo*.

Evidence obtained in different studies up to now suggests that a) ionotropic glutamate receptors of AMPA/kainate type are involved in regulation of OPC proliferation in culture; b) regulation of OPC proliferation *in vivo* may be achieved by glutamatergic signaling at synapses between neurons and OPCs; c) neuronal activity can modulate proliferation of OPCs *in vivo*. However, many questions remain open. In particular, it is unclear whether under normal physiological conditions glutamate inhibits or stimulates proliferation of OPCs. It is also unclear whether glutamate can exert distinct effects on cell proliferation under different conditions, and what are the factors determining whether the effect of glutamate is mitogenic or cytostatic.

OPCs are widespread in the grey and white matter of the CNS and depending on the region examined they comprise 2-9% of the total cell population [1]. Although during development a large proportion of OPCs differentiate into myelinating oligodendrocytes (and perhaps few astrocytes and neurons), many of them stay as OPCs in the adult brain. Adult brain OPCs retain the potential to divide and differentiate into myelinating oligodendrocytes [3, 8, 27], and have a fundamental role in regeneration of myelin after demyelinating diseases of the CNS. Understanding the mechanisms of OPC proliferation and elucidating new possibilities to control their proliferation will greatly contribute to development of new therapeutic strategies for treatment of demyelinating disorders.

Acknowledgments

Our work is supported by the Centre for Integrative Neuroscience (Deutsche Forschungsgemeinschaft, EXC 307). I thank Anna Williams (The University of Edinburgh, UK) for comments on the manuscript.

I declare no conflict of interest.

References

[1] M. R. Dawson, A. Polito, J. M. Levine and R. Reynolds, *Mol. Cell Neurosci.* 24, 476 (2003).
[2] L. Dimou, C. Simon, F. Kirchhoff, H. Takebayashi and M. Gotz, *J. Neurosci.* 28, 10434 (2008).
[3] L. E. Rivers *et al.*, *Nat. Neurosci.* 11, 1392 (2008).
[4] X. Zhu, D. E. Bergles and A. Nishiyama, *Development* 135, 145 (2008).
[5] X. Zhu, R. A. Hill and A. Nishiyama, *Neuron Glia Biol.* 4, 19 (2008).
[6] F. Guo, J. Ma, E. McCauley, P. Bannerman and D. Pleasure, *J. Neurosci.* 29, 7256 (2009).
[7] S. H. Kang, M. Fukaya, J. K. Yang, J. D. Rothstein and D. E. Bergles, *Neuron* 68, 668 (2010).
[8] X. Zhu *et al.*, *Development* 138, 745 (2011).
[9] N. Frohlich, B. Nagy, A. Hovhannisyan and M. Kukley, *J. Anat.* 219, 18 (2011).
[10] D. E. Bergles, J. D. Roberts, P. Somogyi and C. E. Jahr, *Nature* 405, 187 (2000).
[11] R. Chittajallu, A. Aguirre and V. Gallo, *J. Physiol* 561, 109 (2004).
[12] M. Kukley, E. Capetillo-Zarate and D. Dietrich, *Nat. Neurosci.* 10, 311 (2007).
[13] V. Gallo, J. M. Mangin, M. Kukley and D. Dietrich, *J. Physiol* 586, 3767 (2008).
[14] M. Kukley *et al.*, *FASEB J.* 22, 2957 (2008).
[15] M. Kukley, A. Nishiyama and D. Dietrich, *J. Neurosci.* 30, 8320 (2010).
[16] M. Berry, P. Hubbard and A. M. Butt, *J. Neurocytol.* 31, 457 (2002).
[17] R. Karadottir, N. B. Hamilton, Y. Bakiri and D. Attwell, *Nat. Neurosci.* 11, 450 (2008).
[18] L. M. De Biase, A. Nishiyama and D. E. Bergles, *J. Neurosci.* 30, 3600 (2010).
[19] S. C. Lin, D. E. Bergles, *Nat. Neurosci.* 7, 24 (2004).
[20] S. C. Lin *et al.*, *Neuron* 46, 773 (2005).
[21] R. Karadottir, P. Cavelier, L. H. Bergersen and D. Attwell, *Nature* 438, 1162 (2005).
[22] J. L. Ziskin, A. Nishiyama, M. Rubio, M. Fukaya and D. E. Bergles, *Nat. Neurosci.* 10, 321 (2007).
[23] M. R. Dawson, J. M. Levine and R. Reynolds, *J. Neurosci. Res.* 61, 471 (2000).
[24] A. Nishiyama, M. Watanabe, Z. Yang and J. Bu, *J. Neurocytol.* 31, 437 (2002).
[25] P. J. Horner, M. Thallmair and F. H. Gage, *J. Neurocytol.* 31, 469 (2002).
[26] A. A. Aguirre, R. Chittajallu, S. Belachew and V. Gallo, *J. Cell Biol.* 165, 575 (2004).
[27] K. Psachoulia, F. Jamen, K. M. Young and W. D. Richardson, *Neuron Glia Biol.* 5, 57 (2009).

[28] C. Simon, M. Gotz and L. Dimou, *Glia* 59, 869 (2011).
[29] V. Gallo *et al.*, *J. Neurosci.* 16, 2659 (1996).
[30] P. Knutson, C. A. Ghiani, J. M. Zhou, V. Gallo and C. J. McBain, *J. Neurosci.* 17, 2669 (1997).
[31] X. Yuan, A. M. Eisen, C. J. McBain and V. Gallo, *Development* 125, 2901 (1998).
[32] C. Redondo *et al.*, *J. Neurosci. Res.* 85, 1170 (2007).
[33] K. Luyt *et al.*, *J. Neurochem.* 100, 822 (2007).
[34] R. I. Cohen, E. Molina-Holgado and G. Almazan, *Brain Res. Mol. Brain Res.* 43, 193 (1996).
[35] F. Ragheb *et al.*, *J. Neurochem.* 77, 1396 (2001).
[36] F. De Angelis, A. Bernardo, V. Magnaghi, L. Minghetti and A. M. Tata, *Dev. Neurobiol.* (2011).
[37] C. Agresti *et al.*, *Brain Res. Brain Res. Rev.* 48, 157 (2005).
[38] B. Stevens, S. Porta, L. L. Haak, V. Gallo and R. D. Fields, *Neuron* 36, 855 (2002).
[39] M. Wennstrom, J. Hellsten, C. T. Ekdahl and A. Tingstrom, *Biol. Psychiatry* 54, 1015 (2003).
[40] M. Wennstrom, J. Hellsten and A. Tingstrom, *Biol. Psychiatry* 55, 464 (2004).
[41] B. A. Barres, M. C. Raff, *Nature* 361, 258 (1993).
[42] Q. Li, M. Brus-Ramer, J. H. Martin and J. W. McDonald, *Neurosci. Lett.* 479, 128 (2010).
[43] H. Okuda *et al.*, *J. Neurosci. Res.* 87, 3546 (2009).
[44] D. Ehninger *et al.*, *Cell Tissue Res.* 345, 69 (2011).
[45] Y. Matsumoto *et al.*, *PLoS. One.* 6, e27628 (2011).
[46] P. J. Horner *et al.*, *J. Neurosci.* 20, 2218 (2000).
[47] A. Polito, R. Reynolds, *J. Anat.* 207, 707 (2005).
[48] J. Lasiene, A. Matsui, Y. Sawa, F. Wong and P. J. Horner, *Aging Cell* 8, 201 (2009).
[49] G. Alonso, *Glia* 31, 219 (2000).
[50] A. G. Dayer, K. M. Cleaver, T. Abouantoun and H. A. Cameron, *J. Cell Biol.* 168, 415 (2005).
[51] T. Mori, T. Wakabayashi, Y. Takamori, K. Kitaya and H. Yamada, *Acta Histochem. Cytochem.* 42, 1 (2009).
[52] S. Geha *et al.*, *Brain Pathol.* 20, 399 (2010).
[53] A. Nishiyama, X. H. Lin, N. Giese, C. H. Heldin and W. B. Stallcup, *J. Neurosci. Res.* 43, 299 (1996).
[54] M. Velez-Fort, E. Audinat and M. C. Angulo, *Glia* 57, 1104 (2009).
[55] G. Panagiotakos *et al.*, *PLoS. One.* 2, e588 (2007).
[56] N. C. Spitzer, *Nature* 444, 707 (2006).
[57] M. N. Waxham, In: *From Molecules to Networks. An Introduction to Cellular and Molecular Neuroscience* 299, Elsevier Academic Press, (2004).
[58] D. D. Schoepp, P. J. Conn, *Trends Pharmacol. Sci.* 14, 13 (1993).
[59] M. M. Usowicz, V. Gallo and S. G. Cull-Candy, *Nature* 339, 380 (1989).
[60] B. A. Barres, W. J. Koroshetz, K. J. Swartz, L. L. Chun and D. P. Corey, *Neuron* 4, 507 (1990).

[61] D. J. Wyllie, A. Mathie, C. J. Symonds and S. G. Cull-Candy, *J. Physiol* 432, 235 (1991).
[62] D. K. Patneau, P. W. Wright, C. Winters, M. L. Mayer and V. Gallo, *Neuron* 12, 357 (1994).
[63] T. Berger, W. Walz, J. Schnitzer and H. Kettenmann, *J. Neurosci. Res.* 31, 21 (1992).
[64] H. N. Liu, G. Almazan, *Eur. J. Neurosci.* 7, 2355 (1995).
[65] F. Vautier, S. Belachew, R. Chittajallu and V. Gallo, *Glia* 48, 337 (2004).
[66] C. Sato-Bigbee, S. Pal and A. K. Chu, *J. Neurochem.* 72, 139 (1999).
[67] J. R. Johnson, A. K. Chu and C. Sato-Bigbee, *J. Neurochem.* 74, 1409 (2000).
[68] F. S. Afshari, A. K. Chu and C. Sato-Bigbee, *J. Neurosci. Res.* 66, 37 (2001).
[69] M. Pende *et al.*, *J. Neurosci.* 17, 1291 (1997).
[70] C. H. Kastritsis, K. D. McCarthy, *Glia* 8, 106 (1993).
[71] D. R. Marriott, G. P. Wilkin, *J. Neurochem.* 61, 826 (1993).
[72] E. R. Kandel, J. H. Schwartz, T. M. Jessell (Eds.). *Principles of Neural Science.* McGraw-Hill Companies, Inc. (2000).
[73] D. L. Martin, *Glia* 5, 81 (1992).
[74] V. Montana, E. B. Malarkey, C. Verderio, M. Matteoli and V. Parpura, *Glia* 54, 700 (2006).
[75] A. V. Tzingounis, J. I. Wadiche, *Nat. Rev. Neurosci.* 8, 935 (2007).
[76] D. E. Featherstone, S. A. Shippy, *Neuroscientist.* 14, 171 (2008).
[77] M. A. Herman, C. E. Jahr, *J. Neurosci.* 27, 9736 (2007).
[78] M. A. Herman, B. Nahir and C. E. Jahr, *PLoS. One.* 6, e26501 (2011).
[79] J. Muller *et al.*, *J. Gen. Physiol* 134, 115 (2009).
[80] H. R. Parri, T. M. Gould and V. Crunelli, *Eur. J. Neurosci.* 32, 29 (2010).
[81] M. Velez-Fort, P. P. Maldonado, A. M. Butt, E. Audinat and M. C. Angulo, *J. Neurosci.* 30, 6921 (2010).
[82] S. Ge *et al.*, *Nature* 439, 589 (2006).
[83] D. E. Bergles, R. Jabs and C. Steinhauser, *Brain Res. Rev.* 63, 130 (2010).
[84] B. B. Kirby *et al.*, *Nat. Neurosci.* 9, 1506 (2006).
[85] W. P. Ge, W. Zhou, Q. Luo, L. Y. Jan and Y. N. Jan, *Proc. Natl. Acad. Sci. U. S. A* 106, 328 (2009).
[86] S. P. Memberg, A. K. Hall, *J. Neurobiol.* 27, 26 (1995).
[87] I. Ajioka *et al.*, *Cell* 131, 378 (2007).
[88] V. Lobjois, S. Bel-Vialar, F. Trousse and F. Pituello, *Neural Dev.* 3, 4 (2008).
[89] H. V. Friedman, T. Bresler, C. C. Garner and N. E. Ziv, *Neuron* 27, 57 (2000).
[90] Y. Goda, G. W. Davis, *Neuron* 40, 243 (2003).
[91] B. A. Eaton, G. W. Davis, *Genes Dev.* 17, 2075 (2003).
[92] T. S. Otis, Y. C. Wu and L. O. Trussell, *J. Neurosci.* 16, 1634 (1996).
[93] Y. Levchenko-Lambert, D. M. Turetsky and D. K. Patneau, *J. Neurosci.* 31, 9359 (2011).
[94] C. Lappe-Siefke *et al.*, *Nat. Genet.* 33, 366 (2003).
[95] D. Delaunay *et al.*, *J. Neurosci.* 28, 2551 (2008).
[96] J. M. Mangin, V. Gallo, *ASN. Neuro.* 3, e00052 (2011).
[97] B. A. Barres, M. C. Raff, *J. Cell Biol.* 147, 1123 (1999).

[98] B. Czeh *et al.*, *Neuropsychopharmacology* 32, 1490 (2007).
[99] M. Banasr *et al.*, *Biol. Psychiatry* 62, 496 (2007).
[100] A. Trentani, J. Kuipers, G. J. Ter Horst and J. A. den Boer. In: *Handbook of Depression and Anxiety*, 349, Marcel Dekkers, Inc. (2003).
[101] W. P. Ge *et al.*, *Science* 312, 1533 (2006).
[102] G. Y. Wu, K. Deisseroth and R. W. Tsien, *Proc. Natl. Acad. Sci. U. S. A* 98, 2808 (2001).
[103] M. Sheng, G. McFadden and M. E. Greenberg, *Neuron* 4, 571 (1990).
[104] X. Zhang *et al.*, *Proc. Natl. Acad. Sci. U. S. A* 102, 4459 (2005).
[105] A. M. Butt, *Glia* 54, 666 (2006).
[106] F. Kirchhoff, H. Kettenmann, *Eur. J. Neurosci.* 4, 1049 (1992).
[107] S. G. Becker-Catania, J. K. Nelson, S. Olivares, S. J. Chen and G. H. DeVries, *ASN. Neuro.* 3, e00053 (2011).

In: Cell Proliferation
Editors: C. Zhang and X. Zeng
ISBN: 978-1-62417-352-3

Chapter II

The Two Faces of TGF-β in Breast Cancer: Tumour Suppressor and Tumour Promoter

Yihao Li[1], Yvette Drabsch[1] and Peter ten Dijke[1,2,*]
[1]Department of Molecular Cell Biology and Centre for Biomedical Genetics, Leiden University Medical Center, Leiden, the Netherlands and [2]Ludwig Institute for Cancer Research, Uppsala University, Uppsala, Sweden

Abstract

Transforming growth factor-β (TGF-β) has a biphasic role in breast cancer. TGF-β has a cytostatic effect on normal breast epithelial cells and the early phase of breast cancer. During tumour progression, breast cancer cells become insensitive to the growth inhibitory effects of TGF-β. As a result, breast cancer cells start to proliferate faster and produce more TGF-β. By cooperation with oncogenic pathways that abrogate its cytostatic and pro-apoptotic effects, TGF-β then switches from tumour suppressor into tumour promoter. It then stimulates an epithelial to mesenchymal transition of breast cancer cells, which thereby acquire more migratory and invasive properties. Moreover, in a paracrine manner, TGF-β facilitates a favourable microenvironment for rapid tumour growth and metastasis by its immune suppressive and pro-angiogenic effects. Recent studies have shown that TGF-β plays a central role in bone and lung metastasis of breast cancer cells. Therapeutic targeting of this pathway may potentially reduce the invasiveness and metastatic spread of breast cancer cells to distant organs.

* Corresponding author: Peter ten Dijke, Department of Molecular Cell Biology, Leiden University Medical Center, Postal Zone S-1-P, P.O. Box 9600, 2300 RC Leiden, The Netherlands, Email: p.ten_dijke@lumc.nl; Phone: +31 71 526 9200; Fax: + 31 71 526 8270.

List of Abbreviations

ANGPTL4,	angiopoietin-like
AMH,	anti-mullerian hormone
BMP,	bone morphogenetic protein
BRE,	BMP responsive element
CAR,	coxsackie and adenovirus receptor
Co-Smad,	common mediator Smad
CDK,	cyclin dependent kinase inhibitor
CTGF,	connective tissue growth factor
GDF,	growth and differentiation factor
ECM,	extracellular matrix
EMT,	epithelial to mesenchymal transition
Erk,	extracellular signal regulated kinase
FoxH1,	forkhead box protein H1
FoxO,	forkhead box O
Grb,	growth factor receptor binding protein
GS,	glycine and serine rich
GTP,	guanosine-5-triphosphate
HDAC,	histone deacetylase complex
HGF,	hepatocyte growth factor
HIF,	hypoxia-inducible factor
Id,	inhibitor of DNA-binding protein
I-Smad,	inhibitory Smad
IL,	interleukin
JNK,	c-Jun N-terminal kinase
LAP,	latency associated peptide
LIP,	C/EBPβ inhibitory isoform
LTBP,	latent TGF-β binding protein
MAPK,	mitogen-activated protein kinase
MH,	mad homology
Miz,	myc-interacting zinc finger
MMP,	matrix metalloproteinase
MST1,	macrophage stimulating 1
Myc,	myelocytomatosis viral oncogene
P300/CBP,	E1A binding protein p300/CREB-binding protein
PAI,	plasminogen activator inhibitor
pCAF,	p300/CBP-associated factor
PtdIns(3)P,	phosphatidylinositol 3-phosphate
PTHrp,	parathyroid hormone-related protein
R-Smad,	receptor regulated Smad
SARA,	smad anchor for receptor activation
SBE,	smad binding element
Shc,	src homology domain 2 containing adaptor protein
SMA,	smooth muscle actin

Smad,	sma and mad related
Smurf,	smad specific E3 ubiquitin protein ligase
Sos,	son of sevenless
SP1,	specificity protein 1
TAK,	TGF-β activated kinase
TβR,	TGF-β receptor
TGF-β,	transforming growth factor-β
TIMP,	tissue inhibitor of metalloproteinases
TRAF,	tumour necrosis factor receptor associated factor
TSBS,	thrombospondin
VEGF,	vascular endothelial growth factor
ZO-1,	zonula occludens-1

1. Introduction of the TGF-ß Superfamily

The transforming growth factor-β (TGF-β) superfamily is a large group of secreted polypeptides that regulate a plethora of cellular functions, such as cell proliferation, differentiation, adhesion, and cell apoptosis [1].

Since the discovery of its prototypic member in the 1980s, 33 structurally related family members of the TGF-β family have been identified in humans, including TGF-βs, Nodal, bone morphogenetic proteins (BMPs)/growth differentiation factors (GDFs), activins, inhibins and anti-mullerian hormone (AMH) [2–4]. TGF-β family members are highly conserved and found in all multicellular animals [5, 6].

TGF-β family members initiate their cellular responses via a complex of two cell surface receptors with intrinsic serine/threonine kinase activity. Upon ligand-induced receptor activation, the intracellular signalling mediators called Smads (mothers against decapentaplegic homologs) are activated [7].

Smads act as transcription factors that modulate the expression of target genes. Apart from the Smad pathway, there are also non-Smad signalling pathways, which are initiated downstream of the receptor [8, 9].

TGF-β family members induce diverse biological responses. TGF-β is known for its potent growth inhibitory effects in normal and premalignant cells. Tumour cells are frequently resistant to TGF-β-induced cytostatic effects. Moreover, in many cancers it has been shown that TGF-β drives tumour progression by directly affecting migration and invasion of tumour cells [10, 11].

Furthermore, malignant tumour cells frequently secrete high amounts of TGF-β that contribute to a favourable microenvironment by suppressing the immune system and promoting angiogenesis [12, 13]. This chapter reviews the molecular regulation of TGF-β signalling and then summarizes its diverse biphasic roles during breast tumour growth, invasion and metastasis.

2. The Mechanism of TGF-β Signal Transduction

2.1. The TGF-β Pathway

2.1.1. TGF-β is Secreted in a Latent Form

TGF-β family members are multifunctional proteins. Their activities require tight temporal and spatial regulation, which is achieved by producing TGF-β family members in an inactive latent form. TGF-β family precursor proteins are *proteolytically* processed by convertases, producing an N-terminal remnant, also termed a latency-associated peptide (LAP) and a mature, bioactive form [14, 15]. The LAP is folded back onto the mature bioactive part in a non-covalent manner, thereby shielding the mature peptide from receptor binding. The inactive TGF-β complex can be covalently bound to the extracellular matrix (ECM). The latent form can be activated by proteolytic cleavage of the LAP portion, upon which the bioactive part is released. Mature TGF-β is a dimer composed of two 12.5-kd subunits [16]. Matrix metalloproteinase (MMP) 2 and 9, which are anchored on the cell surface via αVβ3 integrin to mediate ECM remodelling, can promote TGF-β liberation [17, 18]. Furthermore, as BMP1 cleavage sites are present in latent TGFβ-binding proteins (LTBPs), BMP1 also cleaves LTBPs to release the large latent complex from the ECM. In the absence of BMP1, the deposition of LAP is induced and the level of activated TGF-β is reduced [19]. Afterwards, biologically activated TGF-β binds to receptors to trigger signalling across the cell membrane.

2.1.2. TGF-β Receptor Activation

Receptor activation begins when the putative mature ligands bind to TGF-β type II receptors (TβRII), which recruit and phosphorylate TGF-β type I receptors (TβRI) [20]. Compared with dozens of ligands among many different species, only five type II receptors and seven type I receptors relay the signals of TGF-β cytokines (Figure 1).

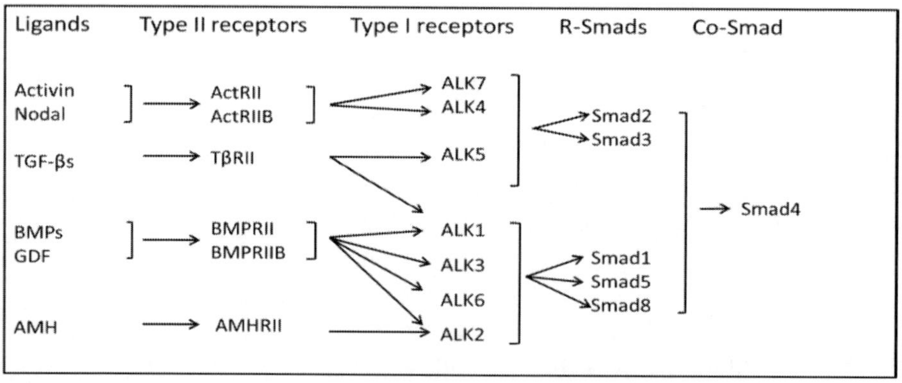

Figure 1. Ligands, receptors, Smads and their interactions. TGF-βs signal via TβRII and ALK5 while Activins and Nodal signal via type II receptors ACTRII, ACTRIIB and type I receptors ALK4 and ALK7. These ligands trigger the phosphorylation of Smad2 and Smad3. BMPs and GDFs bind type II receptors BMPRII, ActRII and ActRIIB. AMH interacts with type II receptor AMHRII. These ligands signal via ALK1, ALK2, ALK3 and/or ALK6 type I receptors, which activate Smad1, Smad5 and Smad8. All R-Smads can associate with Smad4.

Both types of receptors have a ligand-binding region, a single-pass transmembrane domain and a serine/threonine kinase domain [21]. Unlike TβRII, TβRI has an added glycine-serine rich (GS) domain (TTSGSGSG sequence motif) that is important for TβRI activation. In an inactive state, TβRI and TβRII receptors drift on the plasma membrane in dynamic equilibrium.

The immunophilin, FKBP12 (FK506-binding protein) binds to the GS domain of TβRI and blocks receptor activation in the absence of ligands [20]. Upon ligand binding, TβRII recruits TβRI to form the receptor heterotetramer and disrupts the interaction between TβRI and FKBP12, after which TβRII phosphorylates the serine and threonine residues in the GS domain of TβRI [22].

Interestingly, during ligand-receptor complex formation, the binding affinities between the ligands and receptors differ among the various TGF-β subfamilies. The TGF-β and Activin ligands bind with high affinity to their type II receptors without the participation of type I receptors [23, 24]. In contrast, members in the BMP family, such as BMP2 and 4, prefer binding to type I receptors when compared to type II receptors [25].

2.1.3. Smad Activation

As intracellular mediators in TGF-β signalling, Smads were first identified in *Drosophila melanogaster* and *Caenorhabditis elegans* and were named after finding homologues in mammals [26, 27].

There are eight Smads in the human genome, which are usually categorized into 3 subgroups: receptor regulated Smads (R-Smads), the common Smad4 (Co-Smad) and inhibitory Smads (I-Smads) [28, 29]. Five R-Smad transcription factors respond to phosphorylated receptors; Smad1, Smad5 and Smad8 mediate the BMP/GDF/AMH induced signalling pathway, and Smad2, Smad3 act as substrates for TGF-β/Nodal/Activin signalling. Activated R-Smads can form heteromeric complexes with Co-Smad4, which accumulate in the nucleus where they regulate transcriptional responses (Figures 1 and 2). Smad6 and Smad7 are I-Smads that antagonize the activation of R-Smads (see subsequent discussion).

The Smad proteins are comprised of two conserved motifs, the MH (Mad Homology) 1 and 2 domains that are separated by a flexible linker region. R-Smads (but not Smad2) and Smad4 bind to DNA via a hairpin structure in their MH1 domains.

The MH2 domain of R-Smads contains the receptor interaction site-L3 loop (a 17 amino acid region) and a conserved serine-serine-X-serine motif (where X denotes methionine or valine amino acid residues) that is phosphorylated by the activated type I receptor.

In a quiescent state, the R-Smads are predominantly cytoplasmic, and Smad4 is located in both the cytoplasm and nucleus. Upon signal activation, phosphorylated R-Smads and Smad4 accumulate in the nucleus and regulate the transcription of their target genes (Figure 2).

Once the GS domain of TβRI is phosphorylated after ligand binding, R-Smads can be recruited to TβRI by adaptor proteins, such as Smad Anchor for Receptor Activation (SARA; see discussion below) [30]. For the R-Smad-receptor interaction, a nine amino acid sequence between kinase subdomains IV and V in type I receptors (the L45 loop) serves as a key element for R-Smad recruitment [31–33]. The L3 loop in the MH2 domain of R-Smads specifically interacts with the L45 loop in the type I receptors. This interaction can mediate the phosphorylation of serine residues in the carboxy termini of R-Smads. After being released from the receptor complex, the phosphorylated R-Smads display a high affinity for

Smad4. Two activated R-Smads and one Smad4 can form a heteromeric Smad complex, which accumulates in the nucleus [34].

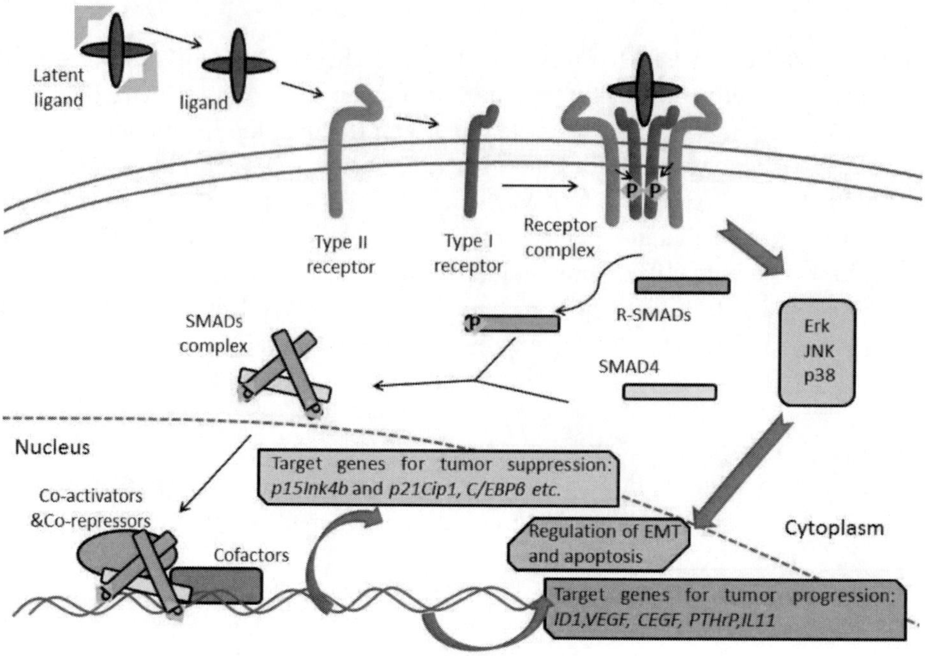

Figure 2. The TGF-β signal transduction pathway. After release from the latent form, mature TGF-β binds to TβRII, and thereafter TβRI is recruited into the complex. The TβRII kinase phosphorylates TβRI, upon which the signal is transmitted into the cell by phosphorylation of R-Smad2 and Smad3. Activated Smad2 and Smad3 can form heteromeric complexes with Smad4, which accumulate in the nucleus. These nuclear Smads function as transcription factors in cooperation with other cofactors, co-activators and co-repressors to regulate gene expression mediating effects on tumour suppression and promotion. Activated TGF-β receptors can also activate Erk and JNK/p38, which can contribute to EMT and apoptosis responses.

2.1.4. Smads as Transcriptional Regulators

Inside the nucleus, the Smad complex initiates transcription of specific genes via a Smad-DNA interaction. DNA recognition by the Smad complex is highly specific. For example, Smad3 and Smad4 selectively bind the sequence of the Smad binding element (SBE), as well as GC-rich nucleotides in the promoters and enhancers of their target genes [35]. In contrast, Smad2 cannot interact with DNA due to a 30 amino acid residue insert near the β hairpin structure in the MH1 domain. Instead, other transcription factors, such as the transcription factor Forkhead box protein H1 (FoxH1), physically interact with the MH2 domain of Smad2 in response to Activin-induced transcription. In fact, a connection between Smads and other DNA-binding transcription factors (co-regulators) is frequently observed; the intrinsic affinity of Smads for DNA is low (Figure 2). Furthermore, Specificity Protein 1 (SP1), a member of the zinc finger family, cooperates with Smad2 and Smad4 to up-regulate TGF-β target genes including *cyclin-dependent kinase 4 inhibitor B p15*, *cyclin-dependent kinase inhibitor 1 p21* and *plasminogen activator inhibitor-1* (*PAI-1*) [36, 37]. This has also been shown in GATA (GATA-binding factors) proteins, where GATA3 acts as a Smad3 binding partner for transcriptional activation, and other transcription factors cooperate with Smad1 to bind the BMP response elements (BREs) in the *Smad7* gene for its transcriptional activation in response to BMP stimulation [38, 39].

2.2. Non-Smad Signalling

Apart from signalling via Smads, TGF-β receptors can also activate intracellular mediators through non-Smad pathways, such as the extracellular signal regulated kinase (ERK) and c-Jun N-terminal kinases (JNK)/p38 mitogen-activated protein kinases (MAPK) pathways [40, 41] (Figure 2). These two atypical signals appear important for certain TGF-β-induced cellular responses, including migration and changes in cell morphology.

Stimulation of Erk phosphorylation by TGF-β has been observed in many cell types [42, 43]. Activation of Erk is required for the epithelial to mesenchymal transition (EMT), which is important for primary tumour cells to become invasive and subsequently for metastasis. TGF-β has been shown to mediate the tyrosine phosphorylation of specific residues on both TGF-β receptors [44]. The Src homology domain 2 containing (Shc) adaptor protein serves as the substrate of tyrosine-phosphorylated receptors [45].

Upon tyrosine phosphorylation Shc can recruit docking proteins, such as the Grb2 (growth factor receptor binding protein 2)/guanine nucleotide exchange factor Son of sevenless (Sos) complex, to stimulate Ras (rapid activation of p21) into its active guanosine-5'-triphosphate (GTP)-bound form [46]. Activated Ras promotes Raf (a MAP kinase kinase kinase) activation, leading to the mitogen-activated protein kinase (MAPK) cascade. The MAPK cascade contains MEK (MEK1 and MEK2) and Erk, which control the transcription of downstream target genes and cellular functions such as cell proliferation, differentiation, migration and cell death [47].

A well-characterised non-Smad pathway involves JNK and p38 MAPK. In the JNK/p38 pathway, JNK and p38 regulate MAPK cascades, which are activated by the MAP kinase kinases (MKKs). TGF-β activates JNK though MKK4, and p38 MAPK though MKK3/6 [48, 49]. For example, studies have shown that TGF-β receptors associate with tumour necrosis factor receptor associated factor (TRAF)6 to induce formation of poly-ubiquitin chains on TRAF6 [50]. TRAF6 then recruits TGF-β-activated kinase 1 (TAK1) to stimulate MKK4 and MKK3/6 [51].

Interestingly, JNK/p38 signalling cooperates with the Smad-dependent pathways to mediate TGF-β-induced cellular responses, such as apoptosis and EMT [52, 53]. The function of non-Smad signalling in breast cancer will be discussed further in this chapter.

2.3. Regulation of the TGF-β-Induced Smad Pathway

To maintain normal cell homeostasis, TGF-β signalling requires precise regulation of both activation and termination. There is a remarkable diversity of co-activators and co-repressors in the signalling network—from the cell membrane to nucleus. Coordinated modulation by regulatory proteins plays important roles in defining cellular responses to TGF-β. This section will describe how the TGF-β pathway is regulated in healthy tissues.

2.3.1. Promotion and Repression of Receptor-Smad Activation

At the receptor level, recruitment of R-Smads requires a group of accessory proteins to mediate type I receptor-induced R-Smad activation.

The anchor protein SARA is localised on early endosomes and binds to phosphatidylinositol 3-phosphate (PtdIns(3)P) through its FYVE finger domain [54]. SARA enhances the interaction between Smad2/3 and the receptor complex upon TGF-β stimulation [55]. Endofin is another FYVE finger-containing protein that is structurally similar to SARA. Endofin efficiently stabilises the complex of BMP receptors and Smad1 and promotes signal transduction [56].

I-Smads inhibit TGF-β/Smad signalling [57] by competing with R-Smad interactions and by recruiting Smad-specific E3 ubiquitin protein ligases (Smurfs) to the activated TGF-β receptor and targeting it for proteosomal degradation. The ubiquitin ligases, Smurf1 and Smurf2 contain WW domains, which bind Smad7 through its PY-motif [58, 59].

2.3.2. Cofactors and Interaction Proteins of Smads

In addition to the regulation at the receptor level, Smads can recruit cofactors and other intracellular proteins for activation, or repression, of target genes. One important cofactor of signalling is the E1A binding protein, p300/CREB-binding protein (p300/CBP), which induces transcriptional activation as a cofactor. To activate gene transcription, Smad4 is required for promoting the interaction between R-Smads and p300/CBP [60]. p300/CBP can promote the acetylation of Smad3 [61]. Another activator called p300/CBP-associated factor (PCAF) enhances the TGF-β induced transcription by using its histone acetyltransferase activity [62].

The oncoproteins, SnoN or Ski, can inhibit TGF-β and BMP signalling via their direct association with the MH2 domain of R-Smads and Smad4. These onocoproteins first form a complex with R-Smads and Smad4 upon stimulation of TGF-β, then they recruit a nuclear corepressor (N-CoR) and histone deacetylase complex (HDAC) to inhibit the Smad-p300/CBP-induced transcription of target genes [63].

3. The Dual Role of TGF-β in Breast Cancer

3.1. Inhibition of Tumourigenesis

In healthy cells, TGF-β maintains cell homeostasis and limits oncogenesis by facilitating growth arrest and apoptosis; however, these functions are deregulated in cancer cells. Precise modulation of the balance between proliferation and apoptosis is critical. In normal tissues, TGF-β is effective in controlling homeostasis and the cellular microenvironment. However, when the TGF-β signalling pathway is perturbed, and cells become (selectively) resistant to its cytostatic effects, hyperplasia and cancer progression can result [64].

3.1.1. Anti-Proliferative and Apoptotic Induction Effects

TGF-β inhibits proliferation in primary mammary epithelial cells through cell-cycle arrest. This anti-proliferative effect is achieved in epithelial cells because TGF-β keeps cells in the G1 phase of the cell cycle. At the transcription level, activated R-Smad complexes combine with Forkhead box O (FoxO) to stimulate the expression of cyclin dependent kinase (CDK) inhibitors, such as *p15Ink4b* and *p21Cip1* [65]. As a result, the complex of cyclinE-cdk4/6 is sequestered by p15Ink4b induction, and cyclinE/A is blocked by expression of p21Cip1. P15 also displaces another CDK inhibitor, p27Kip, to interact with cyclinD-cdk4 and inhibit cyclinE/A-cdk2 [66]. SP1 can bind to Smads and activate the expression of *p21* in the regulation of CDKs [37]. Moreover, the knock down of another TGF-β target gene, *plasminogen activator inhibitor-1* (*PAI-1*), showed interference of TGF-β-induced growth arrest [67].

The oncogene *myelocytomatosis viral oncogene* (Myc) is repressed by TGF-β downstream genes. Upon TGF-β stimulation, the Smad3/4 complex associates with transcription factors E2F4/5, p107 and C/EBPβ to accumulate in the nucleus. Smad3/4 and E2F4/5 can recognize the binding sites of Myc to repress its transcription, and p107 serves as a co-repressor in the TGF-β response. Moreover, C/EBPβ interacts with the promoter of *p15Ink4b* in response to TGF-β, and it is required for repression of *c-myc* [68].

Thus, C/EBPβ plays a key role in TGF-β-mediated cell-cycle arrest [69]. TGF-β prevents the recruitment of Myc to *p15Ink4b* by the myc-interacting zinc finger (Miz). In the absence of Myc, Miz1 upregulates transcription of *p15Ink4b* by binding its promoter [70].

TGF-β inhibits tumour progression directly through cell-autonomous, tumour-suppressive effects. Studies have found the overexpression of Smad3 increased cell death by using the pro-apoptotic functions of the anti-apoptotic genes of BCL families [71]. As described earlier, TGF-β-induced apoptosis is also regulated by non-Smad pathways, such as p38 and JNK via the activation of TAK-1 [72]. TGF-β can also interfere with the activation of the phospho-inositide 3-kinase (PI3K)/Akt/Survivin pathway to induce apoptosis [73].

3.1.2. TGF-β can Induce Breast Tumour Suppression

The gain, and loss, of expression of components in the TGF-β signalling cascade leads to misregulation of signal transduction. This misregulation results in the malignancy and metastasis of breast cancer. Studies have shown that the absence of Smad4 in mouse breast glands led to the transition of mammary epithelial cells into squamous epithelial cells and to the development of squamous cell carcinoma [74]. Interestingly, conditional knockout of TβRII in mammary epithelium was associated with an increased incidence of tumour formation and metastasis in mouse models [75]. Moreover, bioinformatics studies have shown that downregulation of TGF-β receptors and an activator of latent TGF-β, thrombospondin-1 (THBS1), via histone modification, is one of the early steps during breast cancer formation [76]. Furthermore, a xenograft model of breast cancer in nude mice has shown that endogenous TGF-β reduces the size of the putative cancer stem or early progenitor cell population, and secondly it promotes the differentiation of a more committed, but highly proliferative, progenitor cell population to an intrinsically less proliferative state [77].

3.2. Promotion of Tumour Progression

Breast cancer is one of the most common malignant diseases in women. However, it is not the primary tumour, but its metastases at distant sites, that are the main cause of death.

The microenvironment of the primary tumour and the metastatic site are important for the ability of cells to travel to, and take hold, in distant sites. In breast cancers, metastases preferentially invade the bones, lungs, liver, and brain [78]. TGF-β is abundantly expressed in the microenvironment of primary tumours. Breast cancers can selectively shut down the tumour-suppressive arm of the TGF-β signalling pathway and are free to take advantage of its many pro-tumourigenic properties [79].

The basic steps of metastasis include local invasion, intravasation, survival in the circulation, extravasation and finally colonisation [80]. The transition from breast carcinoma *in situ* into invasive carcinoma is poorly understood. Progression to invasion is thought to be promoted by fibroblasts and inhibited by normal myoepithelial cells.

The carcinoma cells can then invade the circulation or lymphatic system. From the lymph nodes, however, there are no direct lymphatic routes to the sites where breast cancer metastases are often found (bone, liver, brain and lungs), so cells in the lymphatic system would eventually need to enter the blood circulation to be transported to these sites, leading to either local tissue invasion or entry into lymph or blood vessels [81]. Carcinoma cells that survive these environments can spread to distant organs. By colonising in adjacent capillaries, tumour cells can extravasate into the microenvironment and form micrometastases. This section will discuss the mechanisms by which TGF-β affects cancer cells and their environment to induce metastasis and growth (Figure 3).

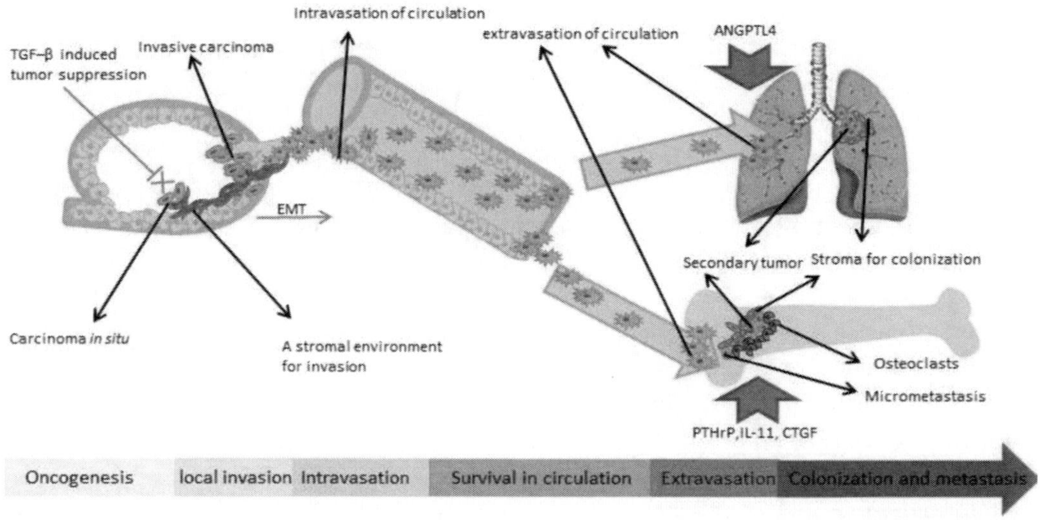

Figure 3. The steps of tumour progression. In the breast gland, the carcinoma *in situ* evades TGF-β-induced tumour suppression and grows towards invasive carcinoma in a supportive stromal environment. Meanwhile, the tumour cells undergo EMT and become mobile and aggressive, which is required for invasion through the endothelial junctions of lymph and blood vessels. After disruption of the nearby endothelial capillary wall, tumour cells circulate in the haematologic system. Then, circulating cells extravasate out of the vasculature and adopt the stroma, which results in homing in the target organs of bone or lung. In bone metastasis, TGF-β induces the expression of osteolytic factors, such as PTHrP, IL-11 and CTGF, to promote osteoclast mobilisation and colonisation. In lung metastasis, ANGPTL4 is stimulated by TGF-β, which is important for metastatic dissemination and colonisation. As a result, micrometastases will be seen in target organs first, which then differentiate into secondary tumours.

3.2.1. Early Stages of Tumour Progression

In the early stages of tumour progression, some cancers evade TGF-β-induced growth arrest by mutating receptors and Smads. Such mutations are rarely seen in breast cancer. The core members of TGF-β signalling remain, and the tumour uses aspects of TGF-β signalling to gain an advantage that enables them to dominate the tissue. For example, downstream anti-growth genes, such as *p15Ink4b* and *C/EBPβ*, can be selectively inhibited. Deletion of the *p15Ink4b* response can increase c-*myc* expression in primary breast cancers. Thus, TGF-β is not acting as a growth inhibitor anymore. Furthermore, studies have shown that C/EBPβ can be blocked by its inhibitory form, called LIP C/EBPβ inhibitory isoform (LIP). High expression of *LIP* will induce a loss of TGF-β-dependent mitotic cycle restriction [69]. Also, *inhibitor of DNA binding protein (Id)1* is usually repressed by TGF-β. However, *Id1* is highly expressed in breast cancer and accelerates the cytostatic and anti-apoptotic functions of carcinoma cells and prevents cell differentiation [82]. Furthermore, Smads can induce *Id1* expression in breast epithelial cells [83].

3.2.2. Angiogenesis

Tumour cells have the ability to make their own blood vessels. Tumour angiogenesis provides nutrients and oxygen required for tumour growth. Furthermore, these vessels can also provide access points for the haematogenous spread of tumour cells [79]. TGF-β can provide a supportive environment to promote angiogenesis. In fact, studies have shown that vascular endothelial growth factor (VEGF) and connective tissue growth factor (CTGF) are direct targets of the TGF-β signalling pathway [84–86]. Hypoxic conditions, present within tumours, can work in conjunction with TGF-β signalling to increase the levels of VEGF mRNA through the activation of hypoxia-inducible factor 1 (HIF1) [86]. HIF-1 and Smad3 also cooperate as a protein complex; they promote the transcription of *VEGF* and, thus, stimulate angiogenesis [85]. As one of the most highly overexpressed genes in bone metastatic populations, the *CTGF* gene product serves as an angiogenic factor, which is induced by the TGF-β/Smad pathway in breast cancer metastasic cells. Silencing Smad3 leads to loss of TGF-β-induced *VEGF* expression in MDA-MB-231 cells, yet a lack of Smad2 increases the levels of *VEGF* [87]. Members of the BMP family also regulate the expression of angiogenic factors or inhibitors in breast tumours. BMP2, 4 and 6 stimulate angiogenesis via stimulation of *VEGF-A* in many cell types [88]. Conversely, high doses of BMP9 repress VEGF-mediated vascularisation by inhibiting basic fibroblast growth factor (bFGF)-stimulated proliferation and migration [89].

3.3. EMT

EMT is a critical mechanism that is mediated by TGF-β, which is frequently observed at the invasive front of the tumour [90, 91]. During breast cancer progression, TGF-β-induced EMT is required for the primary tumour cells to migrate. Breast tumours are commonly epithelial in origin. A loss of epithelial-cell markers and gain of mesenchymal-cell markers has been observed in patient tumour samples, especially at the invasive front of solid tumours. Such changes in epithelial-like and mesenchymal-like cellular markers have been associated with tumour progression and metastatic potential. EMT results from a transcriptional reprogramming of the cells leading to their transition into a mesenchymal-like cellular phenotype. This is promoted by EMT-related signalling pathways driven by abnormal growth factor signals, such as TGF-β. EMT is characterised by loss of cell polarity and epithelial markers (including junctional and cell-cell adhesion proteins, such as Zonula occludens-1 (ZO-1) and E-Cadherin), cytoskeletal reorganisation, and transition to a spindle-shaped morphology concomitant with the acquisition of mesenchymal markers (including α-smooth muscle actin (α-Sma)) and an invasive phenotype. TGF-β is a major inducer of EMT during development, carcinogenesis and fibrosis, with different isoforms mediating various effects depending on the specific cellular context. TGF-β1 was first described as an inducer of EMT in normal mammary epithelial cells. Since this initial study, TGF-β1 has been shown to mediate EMT *in vitro* in a number of different epithelial cells [92]. Smad4 is indispensable for TGF-β-induced EMT. The silencing of Smad4 represses TGF-β-induced EMT and inhibits bone-specific metastases of breast tumours [93]. Smad3 is also required for EMT as it is alleviated in Smad3 knockout mice [94, 95]. Furthermore, two TGF-β transcription factors, Snail and Twist, are regarded as promoters of EMT. Snail interacts with the Smad3/4 complex and downregulates the transcription of *coxsackie* and *adenovirus receptor* (*CAR*), *occludin*, *claudin-3* and *E-cadherin* [96]. Twist promotes EMT by decreasing *E-cadherin* transcription and activating mesenchymal markers [97]. Co-expression of Snail and Smad4 correlated with repression of *CAR* and *E-cadherin* transcription at the invasive fronts of human breast cancers [96]. Recent work has shown that metastasis was disturbed by inhibiting *twist* expression in highly aggressive breast carcinoma cells [97]. Tumour invasion can also involve other cell types to assist with the associated migratory processes. Myofibroblasts have been shown to help tumour cells become more motile by providing growth factors (VEGF) and MMPs [98]. A 3-dimensional model for TGF-β-induced invasion illustrated that Smad3/4 enhances TGF-β-induced invasion by stimulation of MMP-2 and MMP-9 [99]. During tumour invasion, MMPs degrade the extracellular matrix and collagen in stroma [100]. Moreover, MMPs have been shown to directly modulate the secretion of TGF-β and activation of adhesion molecules [101]. Furthermore, TGF-β enhances the activities of MMP-2 and MMP-9 and inhibits the expression of protease inhibitor tissue inhibitor of metalloproteinases (TIMP) in the (breast) tumour, which induces the accumulation of proteinases in the primary microenvironment to promote migration and invasion [102].

3.4. From Primary Site to Circulation

Breast cancer metastasis starts with cell motility in the primary tumour, leading to either local tissue invasion or entry into lymph or blood vessels [103]. Recent work has demonstrated that TGF-β signalling is transiently and locally activated in disseminating single cells, *in vivo*. Blockade of TGF-β signalling prevents cells moving singly *in vivo* but permits cells to move cohesively. Single-cell motility is essential for blood-borne metastasis; whereas, cohesive invasion is capable of lymphatic spread [104].

Tumour imaging has shown that only a small proportion of cancer cells are motile even around the margins of metastatic breast cancer models [105]. This observation suggests that the signalling pathways that promote cancer cell motility may be heterogeneously active in tumours.

Different target genes are involved in modulating the different aspects of cell behavior required for cell motility. Many of these genes could potentially be implicated in the switch between single-cell and cohesive-cell motility. It has been suggested that EGFR, AP-1 family members (c-Jun and JunB), various proteins involved in Rho signalling (RhoC, MPRIP, a Rho-interacting regulator of myosin phosphatase activity, Farp1, a FERM domain-containing Rho exchange factor, Nedd9, an atypical Rac activator and RhoQ/TC10), and a range of molecules implicated in cell–cell adhesion, could all affect the mode of tumour cell motility [104]. After primary breast cancer cells become motile, they need to breech the endothelial barrier to enter the blood stream. The endothelial barrier presents a key rate-limiting step against invasive tumour cells during metastasis [106]. Invasive cancer cells are able to insert pseudopodia through the basement membrane via secretion of bioactively cleaved peptides and degradation of the ECM [107]. A TGF-β-induced protein, Angiopoietin-like 4 (ANGPTL4), could assist cancer cells to disrupt the endothelial capillary wall in primary breast tumours [108].

3.4.1. The Microenvironment for Metastasis

The role of stromal–epithelial crosstalk in the regulation of cancer is important in both tumour suppression and progression. The microenvironment supplies a connective-tissue framework containing ECM, fibroblasts, blood vessel and immune cells [109]. Many studies have shown that there are differences between fibroblast cell populations with respect to regulation by TGF-β [110]. However, it is still not clear what factors are responsible for regulating the differential responses to TGF-β stimulation. It seems logical that there might be a specific molecular profile and microenvironment that is associated with an individual fibroblast population in tissues, which determine the response to TGF-β stimulation *in vivo*. This concept of unique signalling in different fibroblast cell populations has been addressed through global mRNA expression analyses [111]. Work on TGF-β and fibroblasts indicates that TGF-β could function to suppress carcinoma-promoting factors in some fibroblasts, but this mechanism for tumour suppression might be dependent on the distinct molecular profile of each individual fibroblast cell population *in vivo* [112].

The role of TGF-β in tumour promotion might involve the regulation of secreted factors, such as hepatocyte growth factor (HGF), macrophage stimulating 1 (MST1) and TGF-α. These factors are overexpressed by fibroblasts when TβRII expression has been conditionally ablated [112, 113]. Fibroblasts lacking TβRII have been shown to promote the invasion of adjacent carcinoma cells *in vivo*. Furthermore, tumours that were produced from carcinoma cells grafted with fibroblasts that lacked TβRII were more proliferative, had a higher degree of angiogenesis and a lower rate of apoptosis than tumours that were produced from carcinoma cells grafted with normal fibroblasts. In addition, the inhibition of HGF, MST1 or TGF-α signalling in conditioned medium derived from TβRII-deficient fibroblasts has been shown to attenuate the increased proliferation and migration of carcinoma cells treated with the conditioned medium *in vitro*.

Together, these results indicate that distinct TGF-β responses mediated by stromal fibroblasts can regulate carcinoma initiation and progression of adjacent epithelium *in vivo* and *in vitro*. Evidence indicates that the microenvironment is responsible for invasive cells to colonise specific organ types.

3.5. Targeting Specific Organs

In addition to tumour-promoting functions, there is growing experimental evidence that TGF-β can regulate the metastatic process to distinct organs. The evidence from clinical and experimental studies describes a complicated, context-dependent, role for TGF-β in metastasis. The most lucid example is the role of TGF-β in bone metastases. The bone microenvironment has a plentiful supply of growth factors, including TGF-β. The metastatic breast cancer cells that reach this tissue have been shown to release pro-metastatic cytokines that, in turn, activate osteoclast differentiation. Once activated, osteoclasts degrade the bone matrix and release stored TGF-β. Histological analysis demonstrates that 75% of human bone metastasis biopsies show nuclear phosphorylated-Smad2 in the metastatic cells.

These observations indicate functional and active TGF-β signalling in human breast cancer samples. In experimental metastasis assays, invasive breast cancer cells were transduced with retroviral vectors expressing a reporter gene under the control of a TGF-β sensitive promoter. Using this reporter, functional and active TGF-β-Smad signalling was shown specifically in the bone [114]. Expression of Smad7 or a dominant negative TβRII demonstrated that the TGF-β pathway is required for bone metastasis in breast cancer cells [115, 116].

TGF-β promotes breast cancer cells to aggressively metastasise to bone through specific gene inductions. Research has shown that the TGF-β-Smad-signalling pathway induces the production of pro-osteolytic factors, such as parathyroid hormone-related protein (PTHrP) [117]. TGF-β-induced PTHrP stimulates production of the RANK ligand, this in turn enables osteoclast differentiation and promotes bone metastases [118]. Additional factors that may enable TGF-β-mediated bone metastasis include the genes, *interleukin* (*IL*)*11* and *CTGF* (both encode osteolytic factors), which are induced by TGF-β-Smad signalling. By promoting osteoclast functions and further bone degradation, bone metastases set up a vicious cycle wherein TGF-β from the stroma stimulates metastatic cells to activate osteoclasts, which go on to further release TGF-β, thereby perpetuating the bone metastatic lesions. TGF-β can, therefore, exert a pro-metastatic function and facilitate the establishment of metastatic lesions once tumours have reached a secondary site, such as the bone.

TGF-β signalling in the primary tumour may also selectively enhance lung metastasis. To test the requirement of TGF-β signalling for the metastasis of breast cancer to the lung, a breast cancer cell line was used in a xenograft mouse model of metastasis. Abolishing the TGF-β signalling pathway (through expression of a dominant-negative TβRII or through reducing expression of Smad4) reduced the cancer cell's ability to metastasise to the lung from an established primary tumour. In order to understand how the signalling events at a primary tumour are able to contribute to distant metastases, a novel metastatic mechanism was proposed, wherein departing cells are primed by the TGF-β signal to efficiently and specifically colonise the lung. Central to this process is the vascular remodelling gene *ANGPTLngptl4*, which has been identified as a canonical target of TGF-β signalling in multiple breast cancer samples. This gene can disrupt vascular cell-cell junctions and, thus, induce lung vasculature permeability. It was ANGPTLngptl4's vascular remodelling function that was shown to aid cancer cells as they travel through a well-organised vascular barrier like that of the lung [108].

The microenvironment of the bone and lung is vastly different. The bone microenvironment is designed to enable haematopoietic cells to easily shuttle back and forth. The lung environment is organised in sinusoids that contain fenestrated capillary beds [119]. It seems that tumour cells that exhibit enhanced skills at breaching tight vascular barriers would gain a significant advantage in colonising lung while gaining little advantage in colonising bone. On this basis, a new model for TGF-β action in metastasis has been suggested. Unlike previous models, which suggested that TGF-β actions are restricted to the microenvironment, this new model suggests that TGF-β can act at a distance [79]. The cytokine relay between TGF-β and ANGPTLngpt4 enables the actions of TGF-β to project throughout the body, enhancing the reach and impact of TGF-β signalling and metastasis.

Conclusion

As tumours progress into metastasis, there is an increase in cancer-associated morbidity and mortality. Treatments to fight the growth of primary tumours are being developed, while unfortunately, the problem of metastasis is much harder to treat. Clinical and experimental research has provided evidence that TGF-βs are involved in the metastatic process. As such, TGF-β has become a candidate for anti-metastasis therapies. Currently, there are several therapies being tested that effectively target the TGF-β pathway. One of these therapies consists of small-molecule inhibitors that target the receptor kinases. There are also large-molecule neutralising antibodies, and nucleic acid-based therapies. The eventual goal is using these therapies on cancer patients [120, 121]. TGF-β has a complicated role during tumour progression. TGF-β can act as a tumour suppressor or tumour promoter depending on tumour type and stage of tumour progression. Understanding the subtle nature of the TGF-β signalling pathway shows that there is an inherent risk of complications that may arise from using anti-TGF-β therapies. The risks and advantages must be weighted carefully before using these new therapies.

This chapter has focussed on the role of TGF-β in breast cancer. It must be noted that TGF-β's role in tumour growth and eventual metastasis is not limited to breast cancer. Many other cancers, including colon, bladder, lung, and endometrial cancers, as well as sarcomas and melanomas also take advantage of the TGF-β pathway. Some use similar methods as breast cancer to take advantage of TGF-β; other cancers use numerous alternative methods. Regardless, it is hoped that by understanding the mechanisms controlling TGF-β signalling pathways, new tools will be developed to prevent, halt, or treat cancer development, growth and metastasis.

Acknowledgments

Research in our lab on the role of TGF-β in breast cancer is supported by the Netherlands Organization for Scientific Research (NWO 918.66.066), Centre for Biomedical Genetics and Cancerfunden to PtD). Y.L. is supported by a Ph.D. scholarship from the China Scholarship Council. We thank Miriam de Boeck for critical reading of the manuscript. We apologise to those authors that we did not cite.

References

[1] M. Y. Wu, C. S. Hill, *Developmental cell*, 16, 329 (2009).
[2] J. Massagué, *Annual review of biochemistry*, 67, 753 (1998).
[3] G. I. Patterson, R. W. Padgett, *Trends in genetics*, 16, 27 (2000).
[4] R. K. Assoian, A. Komoriya, C. A. Meyers, D. M. Miller, M. B. Sporn, *Journal of Biological Chemistry*, 258, 7155 (1983).
[5] A. Moustakas, S. Souchelnytskyi, C. H. Heldin, *Journal of cell science*, 114, 4359 (2001).
[6] L. Attisano, J. L. Wrana, *Science*, 296, 1646 (2002).
[7] J. Massagué, J. Seoane, D. Wotton, *Genes and development*, 19, 2783 (2005).
[8] Y. E. Zhang, *Cell Research*, 19, 128 (2009).
[9] Y. Mu, S. K. Gudey, M. Landstrom, *Cell and tissue research*, 347, 11 (2012).
[10] W. Cui, D. J. Fowlis, S. Bryson, E. Duffie, H. Ireland, A. Balmain, R. J. Akhurst, *Cell*, 86, 531 (1996).
[11] M. Korpal, Y. Kang, *European journal of cancer*, 46, 1232 (2010).
[12] E. Meulmeester, P. ten Dijke, *The journal of pathology*, 223, 205 (2011).
[13] Y. Drabsch, P. ten Dijke, *Journal of mammary gland biology and neoplasia*, 16, 97 (2011).
[14] C. M. Dubois, M. H. Laprise, F. Blanchette, L. E. Gentry, R. Leduc, *Journal of Biological Chemistry*, 270, 10618 (1995).
[15] D. B. Constam, E. J. Robertson, *The journal of cell biology*, 144, 139 (1999).
[16] B. Schmierer, C. S. Hill, *Nature reviews. Molecular cell biology*, 8, 970 (2007).
[17] Q. Yu, I. Stamenkovic, *Genes and development*, 14, 163 (2000).
[18] N. M. Wolfman, A. C. McPherron, W. N. Pappano, M. V. Davies, K. Song, K. N. Tomkinson, J. F. Wright, L. Zhao, S. M. Sebald, D. S. Greenspan, S. J. Lee, *Proceedings of the National Academy of Sciences of the United States of America*, 100, 15842 (2003).
[19] G. Ge, D. S. Greenspan, *The journal of cell biology*, 175, 111 (2006).
[20] Y. Shi, J. Massagué, *Cell*, 113, 685 (2003).
[21] C. C. Boesen, S. Radaev, S. A. Motyka, A. Patamawenu, P. D. Sun, *Structure*, 10, 913 (2002).
[22] M. Huse, Y. G. Chen, J. Massagué, J. Kuriyan, *Cell*, 96, 425 (1999).
[23] P. J. Hart, S. Deep, A. B. Taylor, Z. Shu, C. S. Hinck, A. P. Hinck, *Nature structural biology*, 9, 203 (2002).
[24] J. Greenwald, M. E. Vega, G. P. Allendorph, W. H. Fischer, W. Vale, S. Choe, *Molecular cell*, 15, 485 (2004).
[25] T. Kirsch, W. Sebald, M. K. Dreyer, *Nature structural biology*, 7, 492 (2000).
[26] J. Raftery, *Journal of evaluation in clinical practice*, 5, 361 (1999).
[27] J. J. Sekelsky, K. S. McKim, G. M. Chin, R. S. Hawley, *Genetics*, 141, 619 (1995).
[28] H. Hayashi, S. Abdollah, Y. Qiu, J. Cai, Y. Y. Xu, B. W. Grinnell, M. A. Richardson, J. N. Topper, M. A. Gimbrone, Jr., J. L. Wrana, D. Falb, *Cell*, 89, 1165 (1997).
[29] A. Nakao, M. Afrakhte, A. Moren, T. Nakayama, J. L. Christian, R. Heuchel, S. Itoh, M. Kawabata, N. E. Heldin, C. H. Heldin, P. ten Dijke, *Nature*, 389, 631 (1997).

[30] G. Wu, Y. G. Chen, B. Ozdamar, C. A. Gyuricza, P. A. Chong, J. L. Wrana, J. Massagué, Y. Shi, *Science*, 287, 92 (2000).
[31] Y. G. Chen, A. Hata, R. S. Lo, D. Wotton, Y. Shi, N. Pavletich, J. Massagué, *Genes and development*, 12, 2144 (1998).
[32] Y. G. Chen, J. Massagué, *Journal of biological chemistry*, 274, 3672 (1999).
[33] X. H. Feng, R. Derynck, *The EMBO journal*, 16, 3912 (1997).
[34] P. ten Dijke, C. S. Hill, *Trends in biochemical sciences*, 29, 265 (2004).
[35] Z. Liu, X. Lin, Z. Cai, Z. Zhang, C. Han, S. Jia, A. Meng, Q. Wang, *Journal of biological chemistry*, 286, 28520 (2011).
[36] X. H. Feng, X. Lin, R. Derynck, *The EMBO journal*, 19, 5178 (2000).
[37] K. Pardali, A. Kurisaki, A. Moren, P. ten Dijke, D. Kardassis, A. Moustakas, *Journal of biological chemistry*, 275, 29244 (2000).
[38] H. Benchabane, J. L. Wrana, *Molecular and cellular biology*, 23, 6646 (2003).
[39] A. Blokzijl, P. ten Dijke, C. F. Ibanez, *Current biology*, 12, 35 (2002).
[40] M. K. Lee, C. Pardoux, M. C. Hall, P. S. Lee, D. Warburton, J. Qing, S. M. Smith, R. Derynck, *The EMBO journal*, 26, 3957 (2007).
[41] R. Derynck, Y. E. Zhang, *Nature*, 425, 577 (2003).
[42] R. S. Frey, K. M. Mulder, *Cancer letters*, 117, 41 (1997).
[43] M. T. Hartsough, K. M. Mulder, *Journal of biological chemistry*, 270, 7117 (1995).
[44] S. Lawler, X. H. Feng, R. H. Chen, E. M. Maruoka, C. W. Turck, I. Griswold-Prenner, R. Derynck, *Journal of biological chemistry*, 272, 14850 (1997).
[45] K. S. Ravichandran, *Oncogene*, 20, 6322 (2001).
[46] A. J. Galliher-Beckley, W. P. Schiemann, *Carcinogenesis*, 29, 244 (2008).
[47] S. Svensson, K. Jirstrom, L. Ryden, G. Roos, S. Emdin, M. C. Ostrowski, G. Landberg, *Oncogene*, 24, 4370 (2005).
[48] C. R. Weston, R. J. Davis, *Current opinion in genetics and development*, 12, 14 (2002).
[49] L. Yu, M. C. Hebert, Y. E. Zhang, *The EMBO journal*, 21, 3749 (2002).
[50] M. Yamashita, K. Fatyol, C. Jin, X. Wang, Z. Liu, Y. E. Zhang, *Molecular cell*, 31, 918 (2008).
[51] A. Sorrentino, N. Thakur, S. Grimsby, A. Marcusson, V. von Bulow, N. Schuster, S. Zhang, C. H. Heldin, M. Landstrom, *Nature cell biology*, 10, 1199 (2008).
[52] L. Xie, B. K. Law, A. M. Chytil, K. A. Brown, M. E. Aakre, H. L. Moses, *Neoplasia*, 6, 603 (2004).
[53] S. Edlund, S. Bu, N. Schuster, P. Aspenstrom, R. Heuchel, N. E. Heldin, P. ten Dijke, C. H. Heldin, M. Landstrom, *Molecular biology of the cell*, 14, 529 (2003).
[54] F. Itoh, N. Divecha, L. Brocks, L. Oomen, H. Janssen, J. Calafat, S. Itoh, P. ten Dijke, *Genes to cells : devoted to molecular and cellular mechanisms*, 7, 321 (2002).
[55] T. Tsukazaki, T. A. Chiang, A. F. Davison, L. Attisano, J. L. Wrana, *Cell*, 95, 779 (1998).
[56] Y. G. Chen, Z. Wang, J. Ma, L. Zhang, Z. Lu, *J. Biol. Chem.*, 282, 9688 (2007).
[57] S. Itoh, P. ten Dijke, *Current opinion in cell biology*, 19, 176 (2007).
[58] T. Ebisawa, M. Fukuchi, G. Murakami, T. Chiba, K. Tanaka, T. Imamura, K. Miyazono, *Journal of biological chemistry*, 276, 12477 (2001).
[59] P. Kavsak, R. K. Rasmussen, C. G. Causing, S. Bonni, H. Zhu, G. H. Thomsen, J. L. Wrana, *Molecular cell*, 6, 1365 (2000).

[60] X. H. Feng, Y. Zhang, R. Y. Wu, R. Derynck, *Genes and development*, 12, 2153 (1998).
[61] Y. Inoue, Y. Itoh, K. Abe, T. Okamoto, H. Daitoku, A. Fukamizu, K. Onozaki, H. Hayashi, *Oncogene*, 26, 500 (2007).
[62] S. Itoh, J. Ericsson, J. Nishikawa, C. H. Heldin, P. ten Dijke, *Nucleic acids research*, 28, 4291 (2000).
[63] Y. Wan, X. Liu, M. W. Kirschner, *Molecular Cell*, 8, 1027 (2001).
[64] J. Massague, *Cell*, 134, 215 (2008).
[65] R. R. Gomis, C. Alarcon, W. He, Q. Wang, J. Seoane, A. Lash, J. Massague, *Proceedings of the National Academy of Sciences of the United States of America*, 103, 12747 (2006).
[66] J. Massagué, S. W. Blain, R. S. Lo, *Cell*, 103, 295 (2000).
[67] R. M. Kortlever, J. H. Nijwening, R. Bernards, *Journal of biological chemistry*, 283, 24308 (2008).
[68] J. Seoane, H. V. Le, L. Shen, S. A. Anderson, J. Massague, *Cell*, 117, 211 (2004).
[69] R. R. Gomis, C. Alarcon, C. Nadal, C. Van Poznak, J. Massagué, *Cancer cell*, 10, 203 (2006).
[70] P. Staller, K. Peukert, A. Kiermaier, J. Seoane, J. Lukas, H. Karsunky, T. Moroy, J. Bartek, J. Massagué, F. Hanel, M. Eilers, *Nature cell biology*, 3, 392 (2001).
[71] J. M. Francis, C. M. Heyworth, E. Spooncer, A. Pierce, T. M. Dexter, A. D. Whetton, *Journal of biological chemistry*, 275, 39137 (2000).
[72] K. Yamaguchi, K. Shirakabe, H. Shibuya, K. Irie, I. Oishi, N. Ueno, T. Taniguchi, E. Nishida, K. Matsumoto, *Science*, 270, 2008 (1995).
[73] J. Wang, L. Yang, J. Yang, K. Kuropatwinski, W. Wang, X. Q. Liu, J. Hauser, M. G. Brattain, *Cancer research*, 68, 3152 (2008).
[74] W. Li, W. Qiao, L. Chen, X. Xu, X. Yang, D. Li, C. Li, S. G. Brodie, M. M. Meguid, L. Hennighausen, C. X. Deng, *Development*, 130, 6143 (2003).
[75] E. Forrester, A. Chytil, B. Bierie, M. Aakre, A. E. Gorska, A. R. Sharif-Afshar, W. J. Muller, H. L. Moses, *Cancer research*, 65, 2296 (2005).
[76] R. A. Hinshelwood, L. I. Huschtscha, J. Melki, C. Stirzaker, A. Abdipranoto, B. Vissel, T. Ravasi, C. A. Wells, D. A. Hume, R. R. Reddel, S. J. Clark, *Cancer research*, 67, 11517 (2007).
[77] B. Tang, N. Yoo, M. Vu, M. Mamura, J. S. Nam, A. Ooshima, Z. Du, P. Y. Desprez, M. R. Anver, A. M. Michalowska, J. Shih, W. T. Parks, L. M. Wakefield, *Cancer research*, 67, 8643 (2007).
[78] D. X. Nguyen, J. Massagué, *Nature reviews genetics*, 8, 341 (2007).
[79] D. Padua, J. Massague, *Cell Res.*, 19, 89 (2009).
[80] D. X. Nguyen, P. D. Bos, J. Massagué, *Nature reviews cancer*, 9, 274 (2009).
[81] A. F. Chambers, A. C. Groom, I. C. MacDonald, *Nature reviews cancer*, 2, 563 (2002).
[82] S. Fong, Y. Itahana, T. Sumida, J. Singh, J. P. Coppe, Y. Liu, P. C. Richards, J. L. Bennington, N. M. Lee, R. J. Debs, P. Y. Desprez, *Proceedings of the National Academy of Sciences of the United States of America*, 100, 13543 (2003).
[83] Y. Y. Liang, F. C. Brunicardi, X. Lin, *Cell Research*, 19, 140 (2009).
[84] L. Pertovaara, A. Kaipainen, T. Mustonen, A. Orpana, N. Ferrara, O. Saksela, K. Alitalo, *Journal of biological chemistry*, 269, 6271 (1994).

[85] T. Sanchez-Elsner, L. M. Botella, B. Velasco, A. Corbi, L. Attisano, C. Bernabeu, *Journal of biological chemistry*, 276, 38527 (2001).
[86] Y. Kang, P. M. Siegel, W. Shu, M. Drobnjak, S. M. Kakonen, C. Cordon-Cardo, T. A. Guise, J. Massague, *Cancer cell*, 3, 537 (2003).
[87] M. Petersen, E. Pardali, G. van der Horst, H. Cheung, C. van den Hoogen, G. van der Pluijm, P. ten Dijke, *Oncogene*, 29, 1351 (2010).
[88] M. M. Deckers, R. L. van Bezooijen, G. van der Horst, J. Hoogendam, C. van Der Bent, S. E. Papapoulos, C. W. Lowik, *Endocrinology*, 143, 1545 (2002).
[89] M. Scharpfenecker, M. van Dinther, Z. Liu, R. L. van Bezooijen, Q. Zhao, L. Pukac, C. W. Lowik, P. ten Dijke, *Journal of cell science*, 120, 964 (2007).
[90] J. Xu, S. Lamouille, R. Derynck, *Cell research*, 19, 156 (2009).
[91] K. Miyazono, *Proceedings of the Japan academy. Series B, physical and biological sciences*, 85, 314 (2009).
[92] B. C. Willis, Z. Borok, *American journal of physiology-lung cellular and molecular physiology*, 293, L525 (2007).
[93] M. Deckers, M. van Dinther, J. Buijs, I. Que, C. Lowik, G. van der Pluijm, P. ten Dijke, *Cancer research*, 66, 2202 (2006).
[94] M. Sato, Y. Muragaki, S. Saika, A. B. Roberts, A. Ooshima, *The journal of clinical investigation*, 112, 1486 (2003).
[95] S. Saika, S. Kono-Saika, Y. Ohnishi, M. Sato, Y. Muragaki, A. Ooshima, K. C. Flanders, J. Yoo, M. Anzano, C. Y. Liu, W. W. Kao, A. B. Roberts, *The American journal of pathology*, 164, 651 (2004).
[96] T. Vincent, E. P. Neve, J. R. Johnson, A. Kukalev, F. Rojo, J. Albanell, K. Pietras, I. Virtanen, L. Philipson, P. L. Leopold, R. G. Crystal, A. G. de Herreros, A. Moustakas, R. F. Pettersson, J. Fuxe, *Nature cell biology*, 11, 943 (2009).
[97] J. Yang, S. A. Mani, J. L. Donaher, S. Ramaswamy, R. A. Itzykson, C. Come, P. Savagner, I. Gitelman, A. Richardson, R. A. Weinberg, *Cell*, 117, 927 (2004).
[98] M. Allinen, R. Beroukhim, L. Cai, C. Brennan, J. Lahti-Domenici, H. Huang, D. Porter, M. Hu, L. Chin, A. Richardson, S. Schnitt, W. R. Sellers, K. Polyak, *Cancer cell*, 6, 17 (2004).
[99] E. Wiercinska, H. P. Naber, E. Pardali, G. van der Pluijm, H. van Dam, P. ten Dijke, *Breast cancer research and treatment*, 128, 657 (2011).
[100] L. A. Liotta, K. Tryggvason, S. Garbisa, I. Hart, C. M. Foltz, S. Shafie, *Nature*, 284, 67 (1980).
[101] O. Tatti, P. Vehvilainen, K. Lehti, J. Keski-Oja, *Experimental cell research*, 314, 2501 (2008).
[102] H. G. Hagedorn, B. E. Bachmeier, A. G. Nerlich, *International journal of oncology*, 18, 669 (2001).
[103] E. Sahai, *Nature reviews cancer*, 7, 737 (2007).
[104] S. Giampieri, C. Manning, S. Hooper, L. Jones, C. S. Hill, E. Sahai, *Nature cell biology*, 11, 1287 (2009).
[105] J. Condeelis, J. E. Segall, *Nature reviews cancer*, 3, 921 (2003).
[106] E. S. Wittchen, J. D. van Buul, K. Burridge, R. A. Worthylake, *Current opinion in hematology*, 12, 14 (2005).
[107] K. Hotary, X. Y. Li, E. Allen, S. L. Stevens, S. J. Weiss, *Genes and development*, 20, 2673 (2006).

[108] D. Padua, X. H. Zhang, Q. Wang, C. Nadal, W. L. Gerald, R. R. Gomis, J. Massague, *Cell*, 133, 66 (2008).

[109] M. M. Mueller, N. E. Fusenig, *Nature reviews cancer*, 4, 839 (2004).

[110] B. Bierie, H. L. Moses, *Nature reviews cancer*, 6, 506 (2006).

[111] H. Y. Chang, J. T. Chi, S. Dudoit, C. Bondre, M. van de Rijn, D. Botstein, P. O. Brown, *Proceedings of the National Academy of Sciences of the United States of America*, 99, 12877 (2002).

[112] N. A. Bhowmick, A. Chytil, D. Plieth, A. E. Gorska, N. Dumont, S. Shappell, M. K. Washington, E. G. Neilson, H. L. Moses, *Science*, 303, 848 (2004).

[113] N. Cheng, N. A. Bhowmick, A. Chytil, A. E. Gorksa, K. A. Brown, R. Muraoka, C. L. Arteaga, E. G. Neilson, S. W. Hayward, H. L. Moses, *Oncogene*, 24, 5053 (2005).

[114] Y. Kang, W. He, S. Tulley, G. P. Gupta, I. Serganova, C. R. Chen, K. Manova-Todorova, R. Blasberg, W. L. Gerald, J. Massague, *Proceedings of the National Academy of Sciences of the United States of America*, 102, 13909 (2005).

[115] J. J. Yin, K. Selander, J. M. Chirgwin, M. Dallas, B. G. Grubbs, R. Wieser, J. Massague, G. R. Mundy, T. A. Guise, *The journal of clinical investigation*, 103, 197 (1999).

[116] D. Javelaud, K. S. Mohammad, C. R. McKenna, P. Fournier, F. Luciani, M. Niewolna, J. Andre, V. Delmas, L. Larue, T. A. Guise, A. Mauviel, *Cancer research*, 67, 2317 (2007).

[117] T. A. Guise, J. J. Yin, S. D. Taylor, Y. Kumagai, M. Dallas, B. F. Boyce, T. Yoneda, G. R. Mundy, *The journal of clinical investigation*, 98, 1544 (1996).

[118] L. A. Kingsley, P. G. Fournier, J. M. Chirgwin, T. A. Guise, *Molecular cancer therapeutics*, 6, 2609 (2007).

[119] Y. Oghiso, O. Matsuoka, *The Japanese journal of experimental medicine*, 49, 223 (1979).

[120] J. M. Yingling, K. L. Blanchard, J. S. Sawyer, *Nature reviews drug discovery*, 3, 1011 (2004).

[121] L. J. Hawinkels, P. ten Dijke, *Growth factors*, 29, 140 (2011).

In: Cell Proliferation
Editors: C. Zhang and X. Zeng
ISBN: 978-1-62417-352-3

Chapter III

MicroRNAs and Their Therapeutic Potential for Vascular Smooth Muscle Cell Proliferation in Restenosis

Eunmi Choi[1,2], *Byeong-Wook Song*[2,3], *Il-Kwon Kim*[4], *Se-Yeon Lee*[2,3], *Min-Ji Cha*[2,3], *Onju Ham*[2,3], *Eunhyun Choi*[5] *and Ki-Chul Hwang*[1,2,3,*]

[1]Severance Biomedical Science Institute
[2]Cardiovascular Research Institute
[3]Brain Korea 21 Project for Medical Science, Yonsei University College of Medicine, Seoul, Republic of Korea
[4]Division of Molecular Cardiology, Department of Medicine, College of Medicine, Texas A and M Health Science Center, Temple, Texas, US
[5]Severance Integrative Research Institute for Cerebral and Cardiovascular Disease, Yonsei University Health System, Seoul, Republic of Korea

Abstract

Percutaneous transluminal coronary angioplasty (PTCA) is a common procedure for treating atherosclerosis, but the long-term efficacy is limited due to the occurrence of restenosis after angioplasty. Restenosis is induced by vascular smooth muscle cell (VSMC) proliferation and migration, inducing the accumulation of extracellular matrix (ECM) and vascular remodeling. The mechanisms of VSMC proliferation may be a potential therapeutic target for the control of restenosis or atherosclerosis. MicroRNAs (miRNAs) are a novel class of non-coding small RNAs (~22 nucleotides) that control

[*] Email address: kchwang@yuhs.ac.

gene expression via the degradation or translational inhibition of their target miRNAs. Each miRNA is able to regulate the expression of multiple target genes, and one target gene can be under the repressive mechanism of multiple miRNAs. These miRNAs are critical regulators of various cellular functions, including proliferation, migration, differentiation and apoptosis. Recent studies have demonstrated that miRNAs also play important roles in the regulation of vascular cell functions, and a number of specific miRNAs are expressed within the vascular system. In response to vascular injury, the expression of miRNAs is dynamically changed. This is especially true for specific miRNAs that are enriched in VSMCs and play a significant role in regulating the proliferation of VSMCs. Furthermore, gain- and loss-of-function miRNA studies have revealed pathogenic and protective functions of miRNAs both in vivo and in vitro. This chapter summarizes the roles of miRNAs in VSMC proliferation with a focus on their therapeutic potential for clinical treatment in vascular disease.

1. Introduction

Atherosclerosis is the primary cause of cardiovascular disease, especially coronary artery disease (CAD), and is the leading cause of morbidity worldwide [1, 2]. Atherosclerosis is a complicated disease triggered by the accumulation of plaque, high plasma concentrations of cholesterol and other factors in the inner arterial wall, stimulating vascular cells to recruit inflammatory molecules, monocytes and T-cells. The migration of smooth muscle cells (SMC) into the intimal portion of the artery and the proliferation and secretion of extracellular matrix proteins induce the formation of fibrous plaque [1, 3-5]. Treatment of atherosclerotic plaques by angioplasty leads to neointima formation, and this remodeling induces restenosis [6, 7] in the significant phenomenon of proliferation-based pathology. Therefore, understanding the mechanism of SMC proliferation and the development of new therapeutic approaches is important for the treatment of atherosclerosis.

MicroRNAs (miRNAs) are small (~22 nucleotides) and non-coding RNA molecules that control gene expression at the post-transcriptional level [8, 9]. miRNA genes are initially transcribed by RNA polymerase II (RNase II) to primary miRNAs (pri-miRNAs), and pri-miRNA is cleaved by Drosha (an RNase III endonuclease) and DGCR8 (a dsRNA binding protein) to precursor miRNA (pre-miRNA), which is then transported to the cytoplasm by Exportin-5 [9, 10]. Pre-miRNA is processed by another RNase III enzyme Dicer to form a double-stranded miRNA duplex. The mature miRNA strand incorporates into the RNA-induced silencing complex (RISC) to negatively affect gene regulation [9]. The miRNA binds complementary 3'untranslated region (3' UTR) in its target mRNA and results in the mediation of translational repression or the degradation of the target mRNA [11, 12]. More than 1,000 human miRNAs have been identified, and it is proposed that they are able to regulate a significant portion of human protein-coding genes [13, 14]. These miRNAs can be located in introns, exons or noncoding genes [15] and are key regulators of various cellular functions, including cell differentiation, proliferation, apoptosis, developmental timing, immunity, stem cell maintenance and pathogenesis [16-18]. Increasing evidence demonstrates that miRNAs regulate cardiovascular disease, particularly in regard to vascular smooth muscle cell (VSMC) proliferation [19-21].

This chapter summarizes the relevance of miRNAs for treating atherosclerosis by controlling VSMC proliferation and the cell cycle involved in restenosis.

2. SMC Proliferation in Restenosis

After angioplasty for treating atherosclerosis, neointima formation and loss of the lumen area are likely to occur; together these phenomena are known as restenosis [6, 7]. A central mechanism of the pathophysiological process of restenosis is known to be inflammation initiated by vascular injury and activated by autocrine or paracrine factors [3]. Restenosis can potentially narrow the diameter of the vessel by more than 50% compared to the reference vessel after angioplasty using a balloon and a stent [20]. Angioplasty is successfully able to prevent artery closure; however, after surgery, the long-term efficacy is limited and can be hallmarked by platelet aggregation, release of growth factors, inflammatory cell infiltration, medial VSMC proliferation and migration, as well as extracellular matrix remodeling [20, 22].

In the early phase of restenosis, initial endothelial bareness and platelet deposition are induced as an immediate response, while VSMC proliferation is seen in the late phase [23]. Forrester *et al.* suggested that there are three phases in restenosis based on the vascular biology of wound healing: an inflammatory phase, a granulation or cellular proliferation phase, and an extracellular matrix remodeling phase [24]. In these phases, stimulatory growth factors (platelet-derived growth factor (PDGF), interleukin-1, interleukin-6 and tumor necrosis factor-α) and chemokines (nitric oxide and heparin sulfate proteoglycan) trigger medial VSMC proliferation and migration within the medial layer. Normal functioning of the endothelial cells is very significant in promoting vasodilatation and suppressing intimal hyperplasia by inhibiting thrombus formation, inflammation, as well as VSMC proliferation and migration. Ohtani *et al.* found that all endothelial cells and some medial SMCs were removed or completely damaged by vascular injury [24]. The enhancement of reendothelialization has a crucial role in inhibiting intimal formation and on the rebuilding of residual endothelial cells. Endothelial cell function is promoted by vascular endothelial growth factor (VEGF), and endothelium suppresses SMC proliferation. Triggering the reendothelialization in the balloon injured area would undermine the development of restenosis. VEGF signaling in particular condition contributes to the diminishment of restenosis by initiating endothelial restoration after vascular injury [25]. However, there are many limitations to study about restenosis in human study. A study of restenosis in animals would be able to directly examine tissue and its effects in various conditions in a way that would not be possible in human studies. Despite the benefits of animal models, they are not perfect for understanding human pathology. Nonetheless, animal studies are the best method for finding answers to biological questions about the mechanisms of human pathology and the pathophysiology of restenosis and for explaining the role of inflammation [23, 26].

3. MiRNAs in SMC Proliferation

Neointimal formation is one of the pathological lesions in various cardiovascular diseases, such as atherosclerosis, post-angioplasty restenosis, coronary heart disease and arterial transplantation. Ji *et al.* studied the aberrant expression of multiple miRNAs in the vascular wall after angioplasty, and 113 of miRNAs were differentially expressed, including 60 up-regulated miRNAs and 53 down-regulated miRNAs. Over time, miR-125a, -125b, -

133, -143, -145 and -365 were seen to be down-regulated in neointimal formation models, while miR-21, -146, -214 and -352 were observed to be up-regulated [27].

In this chapter, the role of miRNAs on SMC proliferation are summarized (Table 1).

Table 1. Summary of microRNAs and their target genes in control of SMC proliferation

Function	microRNA	Target	Reference
↓ proliferation	let-7d	KRAS	[48]
	miR-1	Pim-1	[34]
	miR-133	Sp-1	[36]
	miR-143	Elk-1	[29]
	miR-145	Klf4, Klf5, myocardin	[29,30]
↑proliferation	miR-21	PTEN, Bcl-2, tropomyosin 1, PDCD4, SPRY2, PPARα	[27,39,40]
	miR-26a	SMAD-1, GSK-3β	[41,42]
	miR-208	p21	[45]
	miR-221	c-kit, $p27^{Kip1}$, $p57^{Kip2}$	[32,33]
	miR-222	$p27^{Kip1}$, $p57^{Kip2}$	[32]

3.1. miR-143 and miR-145

Several studies have demonstrated that miR-143 and miR-145 are highly expressed in VSMCs and are related to the regulation of phenotypic changes in VSMCs. miR-143 and miR-145 are highly conserved and lie in close within 1.4 to 1.7Kb on mouse chromosome 18. These miRNAs are enriched in multipotent murine cardiac progenitors before becoming localized to smooth muscle cells [28, 29]. Cordes *et al.* found that the overexpression of miR-145 or miR-143 *in vitro* inhibits the proliferation of VSMCs and is activated by SRF-myocardin interaction. miR-143 and miR-145 were also demonstrated to target several transcription factors, including Kruppel-like factor 4 (Klf4), myocardin and Elk-1, to regulate differentiation and suppress the proliferation of VSMCs [29]. Cheng *et al.* also found that miR-145 modulates the SMC phenotype [30]. In both cultured rat VSMCs *in vitro* and in a rat carotid artery balloon-injury model *in vivo*, miR-145 acted as a novel phenotypic marker and a novel phenotypic modulator of VSMCs. VSMC differentiation marker genes, such as SM α-actin, calponin, and SM-myosin heavy chain (SM-MHC), are up-regulated by miR-145 but are down-regulated by the miR-145 inhibitor. The ability of miR-145 to modify the SMC phenotype occurs through targeting KLF5 gene, which allows this miRNA to increase myocardin mRNA level. Furthermore, the restoration of adenovirus inducing miR-145 in balloon-injured artery model inhibits neointimal growth. Elia *et al.* found that miR-143 and miR-145 have a higher expression in the heart than other organs, and that the expression of these two miRNAs is decreased during acute and chronic vascular stress. The strong expression of miR-143 and miR-145 declined neointimal formation in a rat model of acute vascular injury. This chapter further showed the effect of miRNAs on VSMC differentiation of miR-143 and miR-145 mutant mice. The results showed a significant increase in synthetic VSMCs and a decreased media thickness in the femoral artery and aorta of miR-143 and miR-145 mutant mice [31].

3.2. miR-221 and miR-222

miR-221 and miR-222 are clustered genes and are located in an intergenic region [32]. Deleting miR-221 significantly reduced the induction of PDGF on migration and proliferation which are fundamental characteristics of VSMCs. The expression of miR-221 in VSMCs induced a PDGF signal in injured vascular walls and down-regulated SMC specific genes, such as SM α-actin, smooth muscle calponin and SM22α (SM22). In addition, the down-regulation of SMC-specific genes promoted cell migration and proliferation. The expression of SMC-specific genes modulates c-Kit, and the inhibition of c-Kit by miR-221 resulted in a decrease in SMC differentiation. Moreover, $p27^{Kip1}$ is one of the targets of miR-221; the authors propose that the miR-221-dependent down-regulation of $p27^{Kip1}$ is related to cell proliferation in VSMCs [33].

The targets of miR-221 and miR-222, $p27^{Kip1}$ and $p57^{Kip2}$, promote cell growth and proliferation [32]. This research demonstrated that both miR-221 and miR-222 are critical regulators of VSMCs proliferation and neointimal lesion formation.

3.3. miR-1 and miR-133

miR-1 and miR-133 are cardiac and skeletal muscle-specific and bicistronic transcription miRNAs, which regulate SRF, MyoD, and Mef2. The expression of miR-1 is increased by myocardin, cardiac and smooth muscle-specific transcription co-factors. The up-regulation of miR-1 inhibits Pim-1, an oncogenic serine/threonine kinase, causing SMC proliferation to be down-regulated.

In neointiamal lesions, myocardin and miR-1 are down-regulated, whereas expression level of Pim-1 is increased [34]. It was recently demonstrated that miR-1 regulates the differentiation from embryonic stem cells (ESCs) to SMC by inhibiting KLF4 [35]. miR-133 also negatively regulates PDGF-mediated VSMC proliferation. PDGF treatment inhibits the expression of miR-133 via an extracellular signal-regulated kinase (ERK) signaling pathway. miR-133 targets the transcription factor Sp-1 and then inhibits KLF4 activity, which results in decreased expression of calponin (CNN1), transgelin-2 (TAGLN2) and α-SM actin (ACTA2) and an increase in the expression of SM-MHC (HYH11). After balloon injury *in vivo*, miR-133 was shown to reduce the VSMC phenotype switch and neointimal formation [36].

3.4. miR-21

Reactive oxygen species (ROS) contribute to atherosclerosis and restenosis after angioplasty [37]. miR-21 was significantly expressed in rat carotid arteries after angioplasty and in ROS-induced VSMCs by hydrogen peroxide (H_2O_2) [27, 38] and have a positive effect on VSMC proliferation targeting PTEN and Bcl-2 [27], while having an anti-apoptotic effect on H_2O_2-mediated VSMCs targeting programmed cell death 4 (PDCD4) [38].

The inhibition of miR-21 had a significant negative effect on neointimal lesion formation after angioplasty in a rat carotid artery model. In one set of experiments, the knock-down of miR-21 resulted in decreased cell proliferation and increased cell apoptosis in a dose-dependent manner [27].

In arteriosclerosis obliterans (ASO) associated with hypoxia inducible factor 1-α (HIF-1 α), miR-21 expression was increased about 7-fold, and the proliferation and migration of cultured human arterial SMC were highly decreased by the inhibition of miR-21.

The results showed that tropomyosin 1 is a target of miR-21, as well as the HIF-1α/miR-21/tropomyosin 1 pathway, which possibly has a significant role in the pathogenesis of ASO [39]. Hypoxia up-regulates pulmonary artery SMC proliferation and increases miR-21 expression. miR-21 inhibition is critical in a decrease of hypoxia-induced cell proliferation and migration, but the overexpression of miR-21 in normoxia stimulates cell proliferation. miR-21 down-regulates PDCD4, *Sprouty 2* (SPRY2) and peroxisome proliferator-activated receptor-α (PPARα) protein levels. Through the PPARα 3' UTR luciferase assay, it was found that miR-21 directly targets PPARα [40].

3.5. miR-26a

Leeper *et al.* studied how miR-26a triggers SMC proliferation while inhibiting the differentiation and apoptosis of miR-26a through Transforming growth factor (TGF)-β pathway. SMAD-1 is targeted by miRNA-26a by using luciferase reporter assays. miR-26a was also down-regulated in two mouse models of abdominal aortic aneurysm (AAA) formation, which accelerated its change from a contractile to a synthetic phenotype of SMCs [41].

Airway SMC of Desmin, an intracellular load-bearing protein in null mice, enhanced hypertrophy and miR-26a. Glycogen synthase kinase-3β (GSK-3β) is a target of miR-26a, and the knockdown of miR-26a inhibited airway $SMC^{Des-/-}$ proliferation, as measured by a reduced cell number and DNA synthesis [42].

3.6. miR-208a

miR-208a is encoded by intron 27 of the α-MHC gene and is therefore exclusively expressed in cardiac myocytes [43]. Furthermore, the plasma concentration of miR-208 increases during isoproterenol-induced myocardial injury and shows a good correlation with the plasma concentration of cardiac troponin I (cTnI), a marker of myocardial injury [44].

In insulin-treated VSMC, miR-208 was detect more frequently than a control group and increased the insulin-mediated VSMC proliferation. The role of miR-208 on insulin-mediated VSMC proliferation was attributed to p21 targeting, a key member of the cyclin-dependent kinase (CDK)-inhibitory protein family [45].

3.7. let-7d

Considerable evidence was presented that let-7 regulates cell proliferation in cancer cells by modulating the expression of protein RAS [46, 47]. Yu *et al.* reported that let-7d overexpression resulted in a decrease in KRAS protein and was significantly down-regulated in VSMCs as an important regulator of cell proliferation [48].

4. MiRNAs in Cell Cycle

The cell cycle plays a central role in SMC proliferation for the treatment and prevention of restenotic lesions. Under mitogenic stimuli, the events in the cell cycle are necessary for SMC proliferation and are thus similar or identical to the events necessary for the proliferation of other cell types. Cyclins and CDKs are activated in a sequence which causes cell-cycle events and proliferation. Furthermore, CDK inhibitors (CDKIs) provide negative regulation to this process [49-51]. Activation of CDKs results in the phosphorylation and inactivation of the retinoblastoma protein (Rb). While active Rb is bound to DNA elongation factor (E2F), the inactivation of Rb releases E2F, which allows E2F to initiate the transcription of genes necessary for DNA synthesis, permitting the progression of the cell-cycle. Although various cell types are similar in their cell-cycle control, certain cell-cycle proteins seem to be more involved in SMC proliferation than others. The CDKIs $p27^{Kip1}$ and $p21^{Cip1}$ have been directly implicated in the control of differentiation and proliferation in SMCs, both in vivo and in vitro, compared with the minimal role of $p16^{INK}$ [52]. Over-expression of p21 and p27 resulted in a reduction of neointima formation both in a rat carotid artery model and a pig arterial injury model [53-55].

Many of these miRNAs modify the major proliferation pathways through direct control or targeted components of cell-cycle events. These miRNAs are able to regulate E2F proteins, transcription factors, including key regulators of cell-cycle progression, as well as proliferation [56-58]. Expression of E2F1, as regulator G1 checkpoint to regulate cell cycle progression, is negatively regulated by two miRNAs in this cluster, miR-17-5p and miR-20a [59, 60]. Tumor-suppressive miR-34a is suppressed by cell proliferation through the modification of the E2F signaling pathway (E2F1 and E2F3) in human colon cancer cells [61]. Furthermore, miR-193a significantly reduced cell growth in oral cancer through the modulation of E2F6 [62]. The pRB family inhibits cell-cycle transcription factors of the E2F family and is inhibited by CDK2/4/6, resulting in cell-cycle arrest [53, 63]. In cancer patients, the levels of Rb1 protein are inversely correlated with miR-106a expression. The overexpression of miR-106a is downregulated by pRB [64].

Cyclin/CDK expression is modulated by miRNAs in cell-cycle events. miR-122 inhibits hepatocellular carcinoma cell growth by inducing G2/M arrest, which is associated with reduced cyclin B1 expression by miR-122 [65]. In human embryonic stem cells, miR-195 can downregulate WEE1, a kinase that is a negative regulator of the G2/M modulator CDK1/Cyclin B, and can increase the rate of stem cell division [66]. The protein levels of cyclin D1 and CDK6 are regulated by miR-34a, inducing a significant G1 cell-cycle arrest in the A549 cell line [67]. miR-34 is a direct transcriptional target of p53 and is able to induce cell-cycle arrest by targeting CDK4, Cyclin E and CDK4 levels [68]. Cip/Kip family members (p21, p27 and p57) and INK4a/ARF family members (p14 and p16) are targets of miRNA regulation. miR-221 and miR-222 were found to target p27 and p57 in human carcinomas [69-72]. In addition to cancer cells, PDGF induced miR-221 and the proliferation of VSMCs; downregulation of p27 by miR-221 is critical for PDGF-mediated induction of cell proliferation [33]. p21 is targeted by the miR-17 family miR-106b, and as a result miR-106b alters cell-cycle progression [73, 74]. Interestingly, miR-208 decreased p21 protein expression, but miR-208 overexpression induced VSMC proliferation [45]. Lal *et al.* showed that miR-24 directly represses p16 expression during replicative senescence in human diploid fibroblasts and cervical carcinoma cells [75]. miR-24 inhibition induced senescence or cell-cycle arrest. Also, the silencing of BMI-1, targeted by miR-302, p16INK4a and p14/p19ARF, was further promoted in human pluripotent stem cell proliferation [76]. Numerous lines of evidence implicate miRNAs as major regulators of the cell-cycle and proliferation.

Conclusion

After angioplasty for healing atherosclerosis, restenosis is induced by VSMC accumulation, along with proliferation and migration. Therefore, the mechanisms of VSMC phenotype and response may be potential therapeutic targets for the repair of atherosclerotic lesions. Understanding the molecular mechanisms for VSMC proliferation has led to the development of new therapeutic approaches for patients with coronary artery disease. This chapter focuses on miRNAs as a potential therapeutic strategy by controlling SMC proliferation and cell-cycle events. Several specific miRNAs, including miR-143 and miR-145, miR-221 and miR-222, miR-1 and miR-133, miR-21, miR-26a, miR-208 and let-7d, regulate the proliferation or differentiation of VSMCs and neointimal lesion formation. Although targeting miRNAs represents a potential new therapeutic method to specifically inhibit VSMC proliferation for the prevention of restenosis, future work is required to identify additional miRNAs and their precise targets and to clarify their interactions in vascular disease.

References

[1] A. Curcio, D. Torella and C. Indolfi, *Circ. J.,* 75, 1287 (2011)

[2] A. E. Vendrov, N. R. Madamanchi, Z. S. Hakim, M. Rojas and M. S. Runge, *Circ. Res.,* 98, 1254 (2006).

[3] A. J. Lusis, *Nature*, 407, 233 (2000).

[4] A. Lal, H. H. Kim, K. Abdelmohsen, Y. Kuwano, R. Jr. Pullmann, S. Srikantan, R. Subrahmanyam, J. L. Martindale, X. Yang, F. Ahmed, F. Navarro, D. Dykxhoorn, J. Lieberman and M. Gorospe, *PloS One*, 26, e1864 (2008).

[5] B. N. Davis, A. C. Hilyard, P. H. Nguyen, G. Lagna and A. Hata, *J. Biol. Chem.,* 284, 3728 (2009).

[6] C. J. Sherr, *Cell,* 79, 551 (1994).

[7] C. J. Sherr, *Science,* 274, 1672 (1996).

[8] C. K. Glass and J. L. Witztum, *Cell,* 104, 503 (2001).

[9] C. Xie, H. Huang, X. Sun, Y. Guo, M. Hamblin, R. P. Ritchie, M. T. Garcia-Barrio, J. Zhang and Y. E. Chen, *Stem. Cells Dev.*, 20, 205 (2011).

[10] D. Chen, K. Krasinski, A. Sylvester, J. Chen, P. D. Nisen and V. Andres, *J. Clin. Invest.,* 99, 2334 (1997).

[11] D. P. Bartel, *Cell*, 116, 281 (2004).

[12] D. P. Bartel, *Cell*, 136, 215 (2009).

[13] D. Torella, C. Iaconetti, D. Catalucci, G. M. Ellison, A. Leone, C. D. Waring, A. Bochicchio, C. Vicinanza, I. Aquila, A. Curcio, G. Condorelli and C. Indolfi, *Circ. Res.,* 109, 880 (2011).

[14] E. van Rooij, L. B. Sutherland, X. Qi, J. A. Richardson, J. Hill and E. N. Olson, *Science,* 316, 575 (2007).

[15] F. C. Tanner, Z. Y. Yang, E. Duckers, D. Gordon, G. J. Nabel and E. G. Nabel, *Circ. Res.,* 82, 396 (1998).

[16] F. Fornari, L. Gramantieri, M. Ferracin, A. Veronese, S. Sabbioni, G. A. Calin, G. L. Grazi, C. Giovannini, C. M. Croce, L. Bolondi and M. Negrini, *Oncogene,* 27, 5651 (2008).

[17] F. G. Welt and C. Rogers, Arterioscler. *Thromb. Vasc. Biol.,* 22, 1769 (2002).

[18] F. Sun, H. Fu, Q. Liu, Y. Tie, J. Zhu, R. Xing, Z. Sun and X. Zheng, *FEBS Lett.,* 582, 1564 (2008).

[19] G. K. Hansson and A. Hermansson, *Nat. Immuno.,* 12, 204 (2011).

[20] H. Tazawa, N. Tsuchiya, M. Izumiya and H. Nakagama, *Proc. Natl. Acad. Sci. U.S.A,* 104, 15472 (2007).

[21] I. Bentwich, A. Avniel, Y. Karov, R. Aharonov, S. Gilad, O. Barad, A. Barzilai, P. Einat, U. Einav, E. Meiri, E. Sharon, Y. Spector and Z. Bentwich, *Nat. Genet.,* 37, 766 (2005).

[22] I. Ivanovska, A. S. Ball, R. L. Diaz, J. F. Magnus, M. Kibukawa, J. M. Schelter, S. V. Kobayashi, L. Lim, J. Burchard, A. L. Jackson, P. S. Linsley and M. A. Cleary, *Mol. Cell. Biol.,* 28, 2167 (2008).

[23] I. Shiojima I and K. Walsh, *Circulation,* 110, 2283 (2004).

[24] J. Chen, H. Yin, Y. Jiang, S. K. Radhakrishnan, Z. P. Huang, J. Li, Z. Shi, E. P. Kilsdonk, Y. Gui, D. Z. Wang and X. L. Zheng, *Arterioscler. Thromb. Vasc. Biol.,* 31, 368 (2011).

[25] J. E. 3rd Ferguson and C. Patterson, *Cell Cycle,* 2, 211 (2003).

[26] J. F. O'Sullivan, K. Martin and N. M. Caplice, *J. Am. Coll. Cardiol.,* 57, 383 (2011).

[27] J. K. Gillies and I. A. Lorimer, *Cell Cycle,* 6, 2005 (2007).

[28] J. M. Trimarchi and J. A. Lees, *Nat. Rev. Mol. Cell Biol.,* 3, 11 (2002).

[29] J. Qi, J. Y. Yu, H. R. Shcherbata, J. Mathieu, A. J. Wang, S. Seal, W. Zhou, B. M. Stadler, D. Bourgin, L. Wang, A. Nelson, C. Ware, C. Raymond, L. P. Lim, J. Magnus, I. Ivanovska, R. Diaz, A. Ball, M. A. Cleary and H. Ruohola-Baker, *Cell Cycle,* 8, 3729 (2009).

[30] J. S. Forrester, M. Fishbein, R. Helfant and J. Fagin, *J. Am. Coll. Cardiol.,* 17, 758 (1991).

[31] J. S. Mohamed, A. Hajira, Z. Li, D. Paulin and A. M. Boriek, *J. Biol. Chem.,* 286, 43394 (2011).

[32] J. Sarkar, D. Gou, P. Turaka, E. Viktorova, R. Ramchandran and J.U. Raj, Am. J. Physiol. *Lung Cell. Mol. Physiol.,* 299, L861 (2010).

[33] K. A. O'Donnell, E. A. Wentzel, K. I. Zeller, C. V. Dang and J. T. Mendell, *Nature,* 28, 140 (2009).

[34] K. D. Taganov, M. P. Boldin and D. Baltimore, *Immunity,* 26, 133 (2007).

[35] K. Kozaki, I. Imoto, S. Mogi, K. Omura and J. Inazawa, *Cancer Res.,* 68, 2094 (2008).

[36] K. N. Ivey, A. Muth, J. Arnold, F. W. King, R. F. Yeh, J. E. Fish, E. C. Hsiao, R. J. Schwartz, B. R. Conklin, H. S. Bernstein and D. Srivastava, *Cell stem. cell,* 2, 219 (2008).
[37] K. Ohtani, K. Egashira, K. Hiasa, Q. Zhao, S. Kitamoto, M. Ishibashi, M. Usui, S. Inoue, Y. Yonemitsu, K. Sueishi, M. Sata, M. Shibuya and K. Sunagawa, *Circulation*, 110, 2444 (2004).
[38] K. R. Cordes, N. T. Sheehy, M. P. White, E. C. Berry, S. U. Morton, A. N. Muth, T. H. Lee, J. M. Miano, K. N. Ivey and D. Srivastava, *Nature*, 460, 705 (2009).
[39] L. Elia, M. Quintavalle, J. Zhang, R. Contu, L. Cossu, M. V. Latronico, K. L. Peterson, C. Indolfi, D. Catalucci, J. Chen, S. A. Courtneidge and G. Condorelli, *Cell Death Differ.,* 16, 1590 (2009).
[40] L. He and G. J. Hannon, *Nat. Rev. Genet.,* 5, 522 (2004).
[41] L. He, X. He, L. P. Lim, E. de Stanchina, Z. Xuan, Y. Liang, W. Xue, L. Zender, J. Magnus, D. Ridzon, A. L. Jackson, P. S. Linsley, C. Chen, S. W. Lowe, M. A. Cleary and G. J. Hannon, *Nature,* 447, 1130 (2007).
[42] L. L. Seville, N. Shah, A. D. Westwell and W. C. Chan, Curr. *Cancer Drug Targets,* 5, 159 (2005).
[43] L. S. Lambeth, Y. Yao, L. P. Smith, Y. Zhao and V. Nair, *J. Gen. Virol.* 90, 1164 (2009).
[44] M, Boehm and F. J. Slack, *Cell Cycle,* 5, 837 (2006).
[45] M. A. Costa and D. I. Simon, Circulation, 111, 2257 (2005).
[46] M. L. Yu, J. F. Wang, G. K. Wang, X. H. You, X. X. Zhao, Q. Jing and Y. W. Qin, *Circ. J.,* 75, 703 (2011).
[47] M. R. Kibbe, T. R. Billiar and E. Tzeng, Circ. Res. 86, 829 (2000).
[48] M. T. Pickering, B. M. Stadler and T. F. Kowalik, *Oncogene,* 28, 140 (2009).
[49] M. Wang, W. Li, G. Q. Chang, C. S. Ye, J. S. Ou, X. X. Li, Y. Liu, T. Y. Cheang, X. L. Huang and S. M. Wang, *Arterioscler. Thromb. Vasc. Biol.,* 31, 2044 (2011).
[50] N. J. Leeper, A. Raiesdana, Y. Kojima, H. J. Chun, J. Azuma, L. Maegdefessel, R. K. Kundu, T. Quertermous, P. S. Tsao and J. M. Spin, *J. Cell. Physiol.,* 226, 1035 (2011).
[51] N. Khodayari, K. A. Mohammed, E. P. Goldberg and N. Nasreen, *Cancer Gene Ther.,* 18, 806 (2011).
[52] P. J. Iaquinta and J. A. Lees, Curr. Opin. *Cell Biol.,* 19, 649 (2007).
[53] P. Libby, D. Schwartz, E. Brogi, H. Tanaka and S. K. Clinton, *Circulation*, 86, III47 (1992).
[54] P. Libby, *Nature,* 420, 868 (2002).
[55] R. C. Friedman, K. K. Farh, C. B. Burge and D. P. Bartel, *Genome Res.,* 19, 92 (2009).
[56] R. Ji, Y. Cheng, J. Yue, J. Yang, X. Liu, H. Chen, D. B. Dean and C. Zhang, *Circ. Res.,* 100, 1579 (2007).
[57] R. R. Chivukula and J. T. Mendell, *Trends Biochem. Sci.,* 33, 474 (2008).
[58] R. Ross, *N. Engl. J. Med.,* 340, 115 (1999).
[59] S. Albinsson and W. C. Sessa, *Physiol. Genomics,* 43, 529 (2011).
[60] S. Baskerville and D. P. Bartel, *RNA*, 11, 241 (2005).
[61] S. Galardi, N. Mercatelli, E. Giorda, S. Massalini, G. V. Frajese, S. A. Ciafre and M. G. Farace, *J. Biol. Chem.,* 282, 23716 (2007).
[62] S. L. Lin, D. C. Chang, S. Y. Ying, D. Leu and D. T. Wu, *Cancer Res.,* 70, 9473 (2010).

[63] S. M. Johnson, H. Grosshans, J. Shingara, M. Byrom, R. Jarvis, A. Cheng, E. Labourier, K. L. Reinert, D. Brown and F. J. Slack, *Cell,* 120, 635 (2005).

[64] S. O. Marx, H. Totary-Jain and A. R. Marks, *Circ. Cardiovasc. Interv.,* 4, 104 (2011).

[65] S. Volinia, G. A. Calin, C. G. Liu, S. Ambs, A. Cimmino, F. Petrocca, R. Visone, M. Iorio, C. Roldo, M. Ferracin, R. L. Prueitt, N. Yanaihara, G. Lanza, A. Scarpa, A. Vecchione, M. Negrini, C. C. Harris and C. M. Croce, *Proc. Natl. Acad. Sci. U.S.A.,* 103, 2257 (2006).

[66] T. Kan, F. Sato, T. Ito, N. Matsumura, S. David, Y. Cheng, R. Agarwal, B. C. Paun, Z. Jin, A. V. Olaru, F. M. Selaru, J. P. Hamilton, J. Yang, J. M. Abraham, Y. Mori and S. J. Meltzer, *Gastroenterology,* 136, 1689 (2009).

[67] V. Ambros, *Nature,* 431, 350 (2004).

[68] X. Ji, R. Takahashi, Y. Hiura, G. Hirokawa, Y. Fukushima and N. Iwai, *Clin. Chem.,* 55, 1944 (2009).

[69] X. Liu, Y. Cheng, S. Zhang, Y. Lin, J. Yang and C. Zhang, *Circ. Res.,* 104, 476 (2009).

[70] Y. Cheng, X. Liu, J. Yang, Y. Lin, D. Z. Xu, Q. Lu, E. A. Deitch, Y. Huo, E. S. Delphin and C. Zhang, *Circ. Res.,* 105, 158 (2009).

[71] Y. Lee, M. Kim, J. Han, K. H. Yeom, S. Lee, S. H. Baek and V. N. Kim, *EMBO J.,* 23, 4051 (2004).

[72] Y. Lin, X. Liu, Y. Cheng, J. Yang, Y. Huo and C. Zhang, *J. Biol. Chem.,* 284, 7903 (2009).

[73] Y. Xu, F. Xia, L. Ma, J. Shan, J. Shen, Z. Yang, J. Liu, Y. Cui, X. Bian, P. Bie and C. Qian, *Cancer Lett.,* 310, 160 (2011).

[74] Y. Zhang, Y. Wang, X. Wang, G. M. Eisner, L. D. Asico, P. A. Jose and C. Zeng, *J. Hypertens.,* 29, 1560 (2011).

[75] Y. Zhao and D. Srivastava, *Trends Biochem. Sci.,* 32, 189 (2007).

[76] Z. Y. Yang, R. D. Simari, N. D. Perkins, H. San, D. Gordon, G. J. Nabel and E. G. Nabel, *Proc. Natl. Acad. Sci. U.S.A.,* 93, 7905 (1996).

In: Cell Proliferation
Editors: C. Zhang and X. Zeng

ISBN: 978-1-62417-352-3

Chapter IV

Impaired Proliferation as a Component of the Pathogenesis of Follicular Persistence Associated with Cystic Ovarian Disease

***Natalia R. Salvetti*[1,3], *Florencia Rey*[1,3], *Matías L. Stangaferro*[2], *Eduardo J. Gimeno*[3,4], *Ayelen N. Amweg*[1,3], *Pablo U. Diaz*[1,3] and *Hugo H. Ortega*[1,3*]**

[1]Department of Morphological Sciences
[2]Department of Theriogenology; Faculty of Veterinary Sciences, National University of Litoral, Santa Fe, Argentina
[3]Argentine National Research Council (CONICET)
[4]Institute of Pathology, Faculty of Veterinary Sciences, National University of La Plata, Argentina

Abstract

Folliculogenesis, ovulation, and the subsequent formation of the corpus luteum are complex processes that involve dramatic changes in ovarian cell function. Once initiated, follicular growth is a continuous process without resting phases that ends in ovulation. One of the main modifications in granulosa cell function is the rapid switch from the highly proliferative stage characterizing granulosa cells of preovulatory follicles to the non-proliferative, terminally differentiated phase of luteal cells. Cell cycle regulation involves a balance between several regulatory molecules, which can be altered by numerous external signals in multiple steps. Cystic ovarian disease (COD) and/or polycystic ovarian syndrome (PCOS) are disorders of the reproduction that affect many species, including cattle and human respectively. In the bovine livestock, COD is an

[*] E-mail: hhortega@fcv.unl.edu.ar.

important cause of infertility characterized by anovulation, anestrus, and the persistence of follicles with a diameter larger than that of the ovulatory follicle. The combination of the weak proliferation indices and low apoptosis detected in follicular cysts could explain the cause of the slow growth of cystic follicles and the maintenance of a static condition without degeneration, which leads to their persistence. These alterations might be a result of structural and functional modifications of follicular cells and could be related to hormonal changes in animals with COD.

1. Introduction

During the estrous cycle, the ovary is subjected to extensive tissue remodeling that consists of sequential phases of cell proliferation and cell death in the different ovarian compartments such as follicular, luteal, and interstitial tissues. Follicular development is the central fact, since luteal and interstitial compartments are ultimately derived from growing follicles [1]. Changes are sequentially developed and are dictated by specific, tightly regulated responses to gonadotropins, steroids, and growth factors [2-5]. One of the most important changes occurs in granulosa cell function as a result of a quick modification from the highly proliferative stage characterizing cells of preovulatory follicles to the terminally differentiated phase of luteal cells [5].

Numerous studies have allowed establishing that granulosa cells of primordial follicles are arrested in G0 phase of the cell cycle. These follicles leave this quiescent state and begin to grow gradually [2,5]. However, as these slowly dividing granulosa cells acquire enhanced responsiveness to gonadotropins and begin producing estradiol [6,7], exposure to these hormones triggers a surge of proliferation resulting in the formation of large preovulatory follicles [8].

Granulosa cells of preovulatory follicles not only are highly proliferative but also differentiate and acquire luteinizing hormone (LH) receptors [9]. The LH surge causes dramatic changes in both follicular structure and function. LH terminates follicular growth by causing granulosa cells of preovulatory follicles to exit the cell cycle [8] and rapidly initiates the final differentiation (luteinization) in which the cells stop dividing [10].

Cell cycle progression and proliferation are controlled by a balance of positive and negative regulators converging on cell cycle kinase cascades [11]. Mammalian G1 cyclins and their associated kinases (cyclin-dependent kinases: Cdks) are regulators of the cell cycle that integrate information from outside the cell to lead to G1-phase progression and initiate DNA replication in response to mitogenic signals.

The main components of this family include three D-type cyclins (D1, D2, and D3), which, in different combinations, bind to one of two Cdk subunits, Cdk4 and Cdk6, and the E-type cyclins (E1 and E2), which control the activity of Cdk2 [11,12]. Several combinations of D-type cyclins are expressed in different cell types, whereas cyclin E–Cdk2 complexes are expressed ubiquitously.

Both D- and E-type cyclins, and their associated kinases, are thought to be necessary and limiting for entry into and progression through the G_1 phase [11]. For withdrawal from the cell cycle, proper inactivation of the G1 Cdks, which largely depends on physical association with the Cdk inhibitor proteins, is required. The Cdk inhibitor family consists of the Ink4-type Cdk4/Cdk6 inhibitors, such as p16 (Ink4a), p15 (Ink4b), p18 (Ink4c), and p19 (Ink4d), and the Kip/Cip-type Cdk2 inhibitors, such as p21 (Cip1/Waf1/Sdi1/Cap20), p27 (Kip1), and p57 (Kip2) [13,14].

2. Regulation of the Cell Cycle in the Ovary

Regulation of the cell cycle is a complex mechanism that involves the balance of numerous regulatory molecules, which may be altered by different external signals acting at multiple steps in the cycle. These inputs exist as both soluble extracellular factors, such as growth factors or hormones, and as physical forces of interaction with other cells or with the substratum [13]. In the ovary, estradiol, follicle stimulating hormone (FSH) and LH are essential signals for the growth of preovulatory follicles and their subsequent terminal differentiation as corpora lutea [5,15].

Richards [7] showed that cyclin D2 and p27 exhibit distinctly regulated patterns of expression in granulosa cells of hormone-treated hypophysectomized rats. Specifically, cyclin D2 is expressed at high levels in granulosa cells of preovulatory follicles of hypophysectomized rats treated with estradiol and FSH. When an ovulatory dose of human Chorionic Gonadotropin (hCG) is administered to these rats, cyclin D2 is down-regulated within 4 h and remains low throughout luteinization.

Cyclin D2 is down-regulated specifically in granulosa cells of preovulatory follicles that are destined to ovulate, but continues to be expressed in smaller growing follicles that lack LH receptors [4,5,7]. The p27 Cdk inhibitor is expressed in the preovulatory granulosa cells of hypophysectomized rats treated with estradiol and FSH, and like cyclin D2, is reduced to low levels after 4 h of hCG treatment [4,7].

However, by 24 h after hCG administration, p27 is over-expressed and localized to the luteinizing granulosa cells of the corpus luteum. These hormones also affect the expression of cyclin E, a downstream mediator critical for cell cycle progression through the G1 checkpoint. Low levels of cyclin E are expressed in granulosa cells from ovaries of hypophysectomized rats. *In vivo* treatment with FSH and estradiol increases the expression of cyclin E, whose levels remain high when rats are injected with an ovulatory dose of hCG. However, 48 h after hCG administration, cyclin E levels decrease in luteinized granulosa cells [4,5,7].

Taken together, these data suggest that the mechanism by which the LH surge terminates granulosa cell proliferation involves the rapid inhibition of cyclin D2 transcription. Granulosa cell exit from the cell cycle coincides with the drastic down-regulation of cyclin D2, but prior to the down-regulation of cyclin E and the induction of p27.

Regulation of cyclin D2 is highly probable as the primary regulatory event controlling granulosa cell proliferation [5,16]. Additionally, cyclin E continues to be expressed in luteinizing granulosa cells long after the disappearance of cyclin D2. Therefore, the down-regulation of cyclin D2 in response to LH would presumably prevent the first step in cell cycle progression, thereby initiating granulosa cell exit from the cell cycle before reaching the cyclin E-regulated checkpoint.

The temporal expression pattern for p27 suggests that a second mechanism by which LH terminates granulosa cell proliferation is by increasing the level of this Cdk inhibitor. In addition, the increase in p27 may control some aspects of granulosa cell differentiation or maintenance of luteal cell differentiation [14].

In summary, the LH surge terminates granulosa cell proliferation and initiates differentiation by inverting the balance of positive and negative regulators of cell cycle progression [5,16]. LH coordinately down-regulates the expression of cyclin D2, followed by cyclin E, as it increases the levels of p27. The ovarian phenotypes of mice lacking cyclin D2 or p27 support such a model. The loss of cyclin D2 results in the absence of FSH- and estradiol-stimulated granulosa cell proliferation [16], which normally leads to large, preovulatory follicles.

The loss of p27 results in impaired differentiation, as seen by the inability of granulosa cells to luteinize normally and produce sufficient progesterone to support pregnancy [5,14].

On the other hand, LH also regulates growth and proliferation of theca-interstitial cells. Theca-interstitial cells in the ovary secrete androgens, which are then converted to estrogens by granulosa cells.

The synthesis of androgen by theca-interstitial cells is primarily regulated by LH, which, upon binding to the LH receptor, promotes increased steroid production through the activation of a cAMP-dependent signal transduction cascade.

Palaniappan and Menon [17] observed that the proliferation of theca-interstitial cells increases 6 h after the administration of hCG and that this increase corresponds to an increase in mRNA expression of cyclins D1 and D3. Abnormal function of theca-interstitial cells has been shown to be associated with pathological conditions, such as polycystic ovarian syndrome (PCOS).

3. Ovarian Cysts

Cystic ovarian disease (COD) and/or PCOS are disorders of the reproduction that affect several species of zootechnical interest and humans respectively [18-21]. The heterogeneity of the disease is reflected in numerous animal models of polycystic ovaries (PCO) [22-26].

In the bovine livestock, COD is an important cause of infertility characterized by anovulation, anestrus, and the persistence of follicles with a diameter larger than that of the ovulatory follicle [19]. COD has been defined as the presence of one or more follicular structures in the ovary/ovaries, at least 20 mm in diameter, that persist for more than 10 days in the absence of luteal tissue, interrupting the normal reproductive cycle [19,20,27].

Many factors such as stress, nutritional management and infectious disease can coexist in animals with COD; however, the primary cause has not yet been elucidated. Although it is widely accepted that dysfunction of the hypothalamic-pituitary-gonadal axis is an important etiological factor of COD, delay of follicle regression after ovulation failure is an alternative cause of cysts [28].

Alterations occurring in the ovarian micro-environment of females that present follicular cysts could alter the normal processes of proliferation and programmed cell death in ovarian cells.

4. Altered Proliferation in Follicular Cysts

Studies in experimental models in rats have shown significant differences in the apoptosis and proliferation rates in follicles with induced ovarian cysts. Using a model of induced PCO in rats by exposure to continuous light 24 hours a day for 90 days, Salvetti et al. [29] demonstrated that cellular proliferation, quantified by the immunodetection of PCNA and Ki-67, is higher in the granulosa cells of healthy follicles in the control group than in those of healthy and cystic follicles of light-exposed rats. In theca cells, proliferation is lower in cysts than in tertiary and atretic follicles of all categories presenting scarce proliferation indices with Ki-67.

These results are consistent with those found in cystic follicles induced in rats by estradiol valerate [30]. However, the levels of proliferation in these experimental models differed from those of Das et al. [31], who found high proliferation indices in cystic follicles of women with PCOS, which they attributed to the increased androgen levels detected in womans. It should be considered that androgen levels are not modified in cows with COD [32] or in the light-exposed rat model, and that testosterone levels are even below those found in proestrus [29]. Kafali et al. [33] have recently developed a new animal model for the study of PCO by using letrozole, a non-steroidal aromatase inhibitor. In this model, both testosterone and LH levels are significantly increased, because the conversion of androgen substrates into estrogens is blocked, with the consequent accumulation of androgens. There is also a decrease in estrogen and progesterone and an increase in FSH concentrations [33]. However, even in this experimental model, it could be observed that the levels of proliferation of cystic follicles are lower than those found in growing antral follicles from control animals, in diestrus and proestrus [25].

As described previously, it is well known that granulosa cells proliferate before the cyclic recruitment of follicles under the influence of gonadotropins, achieving high rates of mitosis in the total absence of these hormones. However, the final stage of development prior to ovulation is exclusively dependent on gonadotropins [34,35]. There are reports that indicate that cellular proliferation in the granulosa is regulated by FSH, estrogens and insulin as well as by some growth factors [34,36,37]. Androgens produced by ovarian theca-interstitial cells play a decisive regulatory function in folliculogenesis because they serve as the precursors of estrogen synthesis in granulosa cells. In contrast, estrogens enhance the responsiveness of ovarian follicles to gonadotropin stimulation and increase granulosa cell proliferation. However, excess androgen production impairs follicular function by inhibiting the effect of estrogen on follicular growth, inhibiting FSH induction of LH receptors in granulosa cells [38] and increasing atresia among follicles in rat ovaries [39]. In the normal ovary, estradiol and gonadotropins are essential signals for the growth of preovulatory follicles and their subsequent terminal differentiation as corpora lutea. The LH surge inverts the balance of positive versus negative cell cycle regulators and triggers the granulosa cell exit from the cell cycle concurrently with luteinization [5]. Previous reports have suggested one putative mechanism by which the LH surge terminates granulosa cell proliferation through rapid inhibition of cyclin D2 transcription. This usually happens in each estrous cycle in cattle; however, a variety of hormonal changes that influence the ovarian homeostasis occur in COD.

Several aspects of the disease, including the changes occurring at the hypothalamic-pituitary level and those occurring locally in the ovary, have been examined in cattle. Most of these investigations have been carried out in samples obtained at slaughterhouse, and this material have shown a lower index of proliferation in all follicular layers of ovarian cysts in cattle, with an intense proliferation in the basal area of the granulosa layer of normal tertiary follicles and a decrease in atretic and cystic follicles [40,41].

Ovarian cysts have been induced in cattle by a variety of treatments given in late diestrus or proestrus such as administration of estrogens, a combination of progesterone and estradiol, antiserum against bovine LH, testosterone or exogenous adrenocorticotropic hormone (ACTH) [42]. ACTH suppresses baseline LH values during the follicular phase of the estrous cycle [42] and it has been demonstrated that it delays the LH surge by a direct effect on the pituitary gland [43]. Alterations in basal secretions of LH modulate follicular development during the estrous cycle [44]. Lengthening the luteal phase with low levels of progesterone (1–2 ng/ml) results in prolonged development of follicles, which indicates that subtle changes in the hormonal milieu can dramatically alter the normal pattern of follicular development in cattle [45].

Ribadu et al. [42] showed that LH pulse frequency is reduced during cyst formation and persistence in the ACTH induction model, which suggests that alterations in the pulsatile release of LH might cause ovarian follicular cyst formation in heifers. Changes in the normal pattern of LH secretion affect the presence of LH receptors and steroid hormone receptors. LH receptor mRNA is increased in the follicles of cows with COD [46], whereas estrogen receptor beta is decreased in follicles of cows with COD [47]. Several growth factors such as insulin-like growth factor-1 (IGF1) also promote granulosa cell survival [48]. In ovarian cysts, an altered amount of IGF1 protein probably related to reduced cell proliferation has been observed in these structures [49,50]. This information indicates that hormonal changes result in altered amounts of proteins involved in the cell cycle such as cyclins D1, D2 and E, which contribute to follicular persistence. Our results in ACTH-induced cystic ovaries indicate a clear decrease in the index of proliferation in all cellular layers of cysts, similar to that observed in atretic follicles. Also, mRNA for D1 and E cyclins in cystic follicles is decreased as compared with that in healthy tertiary follicles [28]. In the same study, we found significant alterations in cell proliferation and apoptosis rates in follicles from cows with induced ovarian cysts (Figures 1 and 2).

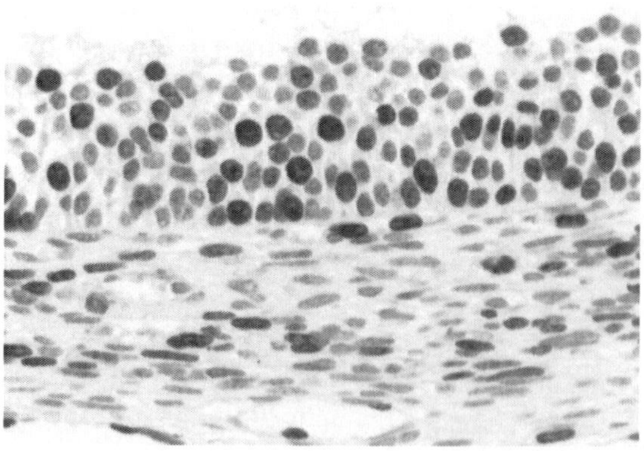

Figure 1. Follicular wall of healthy antral follicle from a normal bovine ovary stained with anti-ki67. X40.

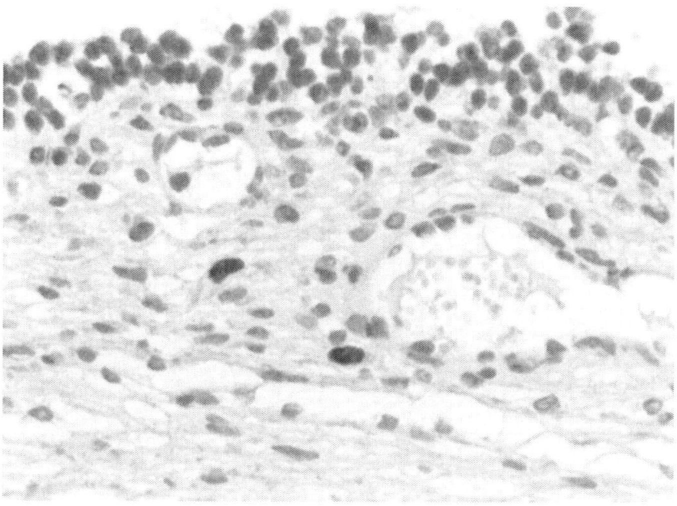

Figure 2. Follicular wall of a follicular cyst from a cow with COD stained with anti-ki67. X40.

These findings support the notion that follicular persistence is an important component of COD pathogenesis. These results are consistent with those found in induced cystic follicles in rats in different experimental models [25,29,30].

5. Decreased Apoptosis Contributes to Follicular Persistence

An adequate balance between survival and apoptotic factors may determine whether a follicle will either continue developing or undergo atresia [51-53]. The progression of apoptosis in follicular cells depends on the cooperative regulation of different paracrine and autocrine factors. Regulation of apoptotic signaling in the ovary is generally achieved by the Fas system and the B-Cell Lymphoma-2 (Bcl-2) family [52,54,55]. Members of the Bcl-2 family of proteins are considered among the main regulatory proteins acting at the mitochondrial level. Considering their function, Bcl-2 members can be divided into antiapoptotic (Bcl-2, Bcl-W, Bcl-xL) and proapoptotic (Bax, Bad, Bim, Bcl-xS, Bod, Bok/Mtd) agents. Antiapoptotic proteins are able to block the activation of effector caspases, caspase-3, caspase-6 and caspase-7, which in turn transduce the apoptotic signals [56,57]. Bcl-2 resides on the nuclear/endoplasmic reticulum membrane, with a smaller portion on the mitochondrial membrane. It has been suggested that when Bcl-2 is present in the outer mitochondrial membrane it can block apoptosis by inhibiting the release of apoptosis-inducing factors, cytochrome c and the intermembrane protein DIABLO-Smac from mitochondria [57,58]. Bcl-xL and Bcl-w are usually abundant in the mitochondrial membrane where they inhibit the release of apoptosis-inducing factors [52]. On the other hand, the proapoptotic protein Bax plays a major role in initiating the release of cytochrome c. This protein may form a pore in the outer membrane of mitochondria, allowing cytochrome c to leak out. Therefore, the Bax:Bcl-2 (or another antiapoptotic protein such as Bcl-xL or Bcl-w) ratio might be important in the mitochondria-dependent apoptosis cascade for the release of cytochrome c [57,58].

In an experimental model in rats, we have demonstrated that DNA fragmentation, as well as activated caspase-3 and Bax protein expression, are significantly lower in all cell layers of tertiary and cystic follicles from COD rats than in normal atretic follicles from PCO and control groups, and that the expression of survival proteins of the Bcl-2 family such as Bcl-2, Bcl-xL and Bcl-w is high in healthy and cystic follicles in both groups [29].

Anderson and Lee [59] found that during cystogenesis in a dehydroepiandrosterone (DHEA) induction model, apoptosis systematically progresses from the cumulus towards the mural granulosa layer and that the outer layer of mural granulosa cells escapes apoptosis. In contrast, granulosa cells of atretic follicles undergo apoptosis in a random manner. Isobe and Yoshimura [40,41] found low proliferation and apoptosis in cystic follicles in cows when compared with healthy follicles.

In ACTH-induced COD in bovines, DNA fragmentation, caspase-3, Fas ligand (FASLG) and Bax protein are significantly lower in layers of tertiary and cystic follicles from COD cows than in normal atretic follicles of both groups, whereas Bcl-2 expression is higher in growing and cystic follicles in both groups (Figures 3 and 4).

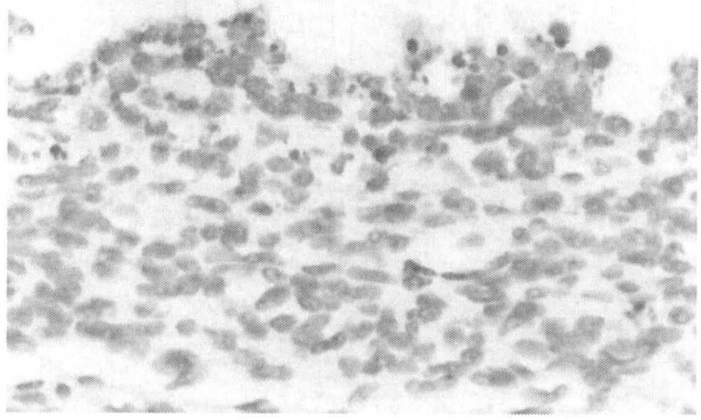

Figure 3. Follicular wall of an atretic follicle from a normal bovine ovary showing *in situ* apoptosis by TUNEL. X40.

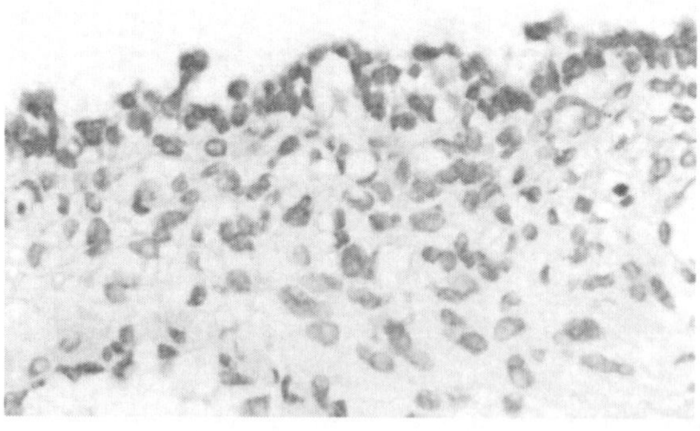

Figure 4. Follicular wall of a cyst from a cow with COD showing *in situ* apoptosis by TUNEL. X40.

No differences were found in the TNF receptor superfamily member 6 (Fas) in atretic and cystic follicles, being higher than in secondary and tertiary follicles. In our previous studies, we found that the Bax and Bcl-2 proteins were coincident with the gene expression evaluations, being similar in tertiary and cystic follicles, and that there was a relatively greater Bcl-2 than Bax in cystic follicles as compared to tertiary follicles, as indicated by multiplex analysis [28]. In studies carried out in women with PCOD, Das et al. [31] found high expression of antiapoptotic factors such as cellular inhibitor of apoptosis 2 (cIAP-2) and Bcl-xL and lower expression of Bax and Caspase-3 in granulosa cells of cystic follicles, with values similar to those found in healthy follicles. On the other hand, Almahbobi et al. [60] found that granulosa cells from ovarian cysts of women with PCOS are normal, with low levels of apoptosis and high expression of gonadotropin receptors, specifically FSH receptor (FSHr).

The response of the ovarian follicle to the combined effect of survival and death factors determines its ultimate fate: follicular atresia or ovulation. Progesterone, one of the factors induced by LH, has been reported to act as an antiapoptotic factor in luteinized rat and human granulosa cells [61]. Tropic hormones important for cell proliferation also have a role in the suppression of apoptosis or cell survival. FSH and LH are important factors involved in the proliferation of follicular somatic cells and development of preovulatory follicles. Gonadotropins have also been demonstrated to affect the apoptotic machinery by suppressing the expression of proapoptotic proteins [56,62] as well as by inducing the expression of antiapoptotic proteins [61]. However, Yacobi et al. [61,63] demonstrated that although gonadotropins (mainly LH) decrease apoptosis in granulosa cells in cultured rat preovulatory follicles, they increase apoptosis in theca/interstitial cells through the caspase-3 cascade. Tilly [56] demonstrated an inhibition of granulosa cell apoptosis and follicular atresia mediated by gonadotropin treatment, which may be associated with the ability of gonadotropins to reduce the amount of Bax present in granulosa cells, while maintaining a constitutive level of Bcl-2 and Bcl-xL expression. Moreover, the level of Bcl-xS mRNA is reduced by gonadotropin treatment and this effect may contribute to the shift in the balance between death inducer and death repressor gene expression. It is also possible that other hormonal signals such as ovarian steroids or locally produced growth factors, potential modulators of granulosa cells [48], serve as the primary regulators of Bcl-2 and Bcl-xL gene expression.

The data presented here indicate that the cells of the follicular cysts are functional and, in fact, several authors have demonstrated this functionality, analyzing cellular production of hormones, enzymes and their response to different stimuli [60,64,65]. In cattle, during the first follicular wave of the estrous cycle, expression of FAS and FASLG genes is increased in subordinate follicles compared with dominant follicles [66,67]. *In vitro* studies [68] indicate that granulosa cells possess endogenous pathways to activate apoptosis that are inhibited in the presence of survival factors. For example, granulosa cells of cattle express the FAS gene but are resistant to death by exogenous FASLG *in vitro* when serum is present in the culture medium [67-70]. The increased availability of survival factors such as IGF in the dominant follicles of cattle *in vivo* inhibits the expression of the FAS and FASLG genes and prevents activation of the FAS pathway. The effect of growth factors such as IGF1 to protect cells from apoptosis seems to be correlated with their ability to stimulate progression through the cell cycle. IGF1, basic fibroblast growth factor and epidermal growth factor decrease the FASLG-induced apoptosis of cultured granulosa cells of cattle and simultaneously increase cell proliferation [68]. Follicular cysts have greater amounts of IGF2, a growth factor that acts to prevent cellular death [50].

Conclusion

Although much remains to be done in order to characterize the pathogenesis of cystic ovaries, we can confirm that cellular proliferation and apoptosis are altered in cystic follicles. The combination of the weak proliferative activity and low levels of apoptosis observed in follicular cysts could explain why these follicles grow slowly and then maintain a static condition without degeneration, which leads to their persistence. These alterations may be due to structural and functional modifications that take place in follicular cells and could be related to the hormonal changes that occur in animals with COD.

References

[1] P. K. Pradeep, X. Li, H. Peegel and K. M. Menon, *Endocrinology*. 143, 2930 (2002).
[2] N. Hirshfield, *Int. Rev. Cytol.* 124, 43 (1991).
[3] T. Pederson, *Acta. Endocrinol.* 64, 304 (1970).
[4] J. S. Richards, *The Vertebrate Ovary: Comparative Biology and Evolution*. Plenum Press, New York (1978)
[5] R. L. Robker and J. S. Richards, *Biol. Reprod.* 59, 476 (1998).
[6] J. S. Richards, *Endocrinology*. 97, 1174 (1975).
[7] J. S. Richards, *Physiol Rev.* 60, 51 (1980).
[8] M. C. Rao, A. R. Midgley Jr. and J. S. Richards, *Cell*. 14, 71 (1978).
[9] J. Th. J. Uilenbroek and J. S. Richards, *Biol. Reprod.* 20, 1159 (1979).
[10] J. S. Richards, *Endocr. Rev.* 15, 725 (1994).
[11] C. J. Sherr and J. M. Roberts, *Genes Dev.* 18, 2699 (2004).
[12] D. O. Morgan, *Annu. Rev. Cell Dev. Biol.* 13, 261 (1997).
[13] Vidal and A. Koff, *Gene*. 247, 1 (2000).
[14] S. Jirawatnotai, D. S. Moons, C. O. Stocco, R. Franks, D. B. Hales, G. Gibori and H. Kiyokawa, *J. Biol. Chem.* 278, 17021 (2003).
[15] J. S. Richards and H. J. Kirchick, *Biol. Reprod.* 30, 737 (1984).
[16] P. Sicinski, J. L. Donaher, Y. Geng, S. B. Parker, H. Gardner, M. Y. Park, R. L. Robker, J. S. Richards, L. K. McGinnis, J. D. Biggers, J. J. Eppig, R. T. Bronson, S. J. Elledge and R. A. Weinberg, *Nature*, 384, 470 (1996).
[17] M. Palaniappan and K. M. Menon, *Mol. Endocrinol.* 24, 1782 (2010).
[18] J. Jakimiuk, S. R. Weitsman, H. W. Yen, M. Bogusiewicz and D. A. Magoffin, *J. Clin. Endocrinol. Metab.* 87, 5532 (2002).
[19] W. J. Silvia, T. B. Hatler, A. M. Nugent and L. F. Laranja Da Fonseca, *Domest. Anim. Endocrinol.* 23, 167 (2002).
[20] T. Vanholder, G. Opsomer and A. de Kruif, *Reprod Nutr Dev.* 46, 105 (2006).
[21] Y. L. Sun, J. Zhang, Z. G. Ping, C. Q. Wang, Y. F. Sun, L. Chen, X. Y. Li, C. J. Li, X. L. Zhu, Z. Liu, W. Zhang and X. Zhou, *Reprod. Domest. Anim.* doi: 10.1111/j.1439-0531.2011.01929.x. (2011).
[22] V. B. Mahesh, T. M. Mills, C. A. Bagnell and B. A. Conway, *Adv. Exp. Med. Biol.* 219, 237 (1987).
[23] D. K. Mahajan, *Endocrinol. Metab. Clin. North. Am.* 17, 705 (1988).

[24] N. R. Salvetti, A. M. Canal, E. J. Gimeno and H. H. Ortega, *Braz. J. Vet. Res. Anim. Sci.* 41, 389 (2004).
[25] C. Baravalle, N. R. Salvetti, G. A. Mira, N. Pezzone and H. H. Ortega, *Arch Med Res.* 37, 830 (2006).
[26] H. H. Ortega, N. R. Salvetti and V. Padmanabhan, *Reproduction.* 137, 865 (2009).
[27] T. Peter, *Reprod. Domest. Anim.* 39, 1 (2004).
[28] N. R. Salvetti, M. L. Stangaferro, M. M. Palomar, N. S. Alfaro, F. Rey, E. J. Gimeno and H. H. Ortega, *Anim. Reprod. Sci.* 122, 98 (2010).
[29] N. R. Salvetti, C. G. Panzani, E. J. Gimeno, L. G. Neme, N. S. Alfaro and H. H. Ortega. *Reprod. Biol. Endocrinol.* 7, 68 (2009).
[30] C. Baravalle, N. R. Salvetti, G. A. Mira, J. A. Lorente and H. H. Ortega, *Physiological Research.* 56, 67 (2007).
[31] M. Das, O. Djahanbakhch, B. Hacihanefioglu, E. Saridogan, M. Ikram, L. Ghali, M. Raveendran and A. Storey, *J. Clin. Endocr. Metab.* 93, 881 (2008).
[32] H. S. Choi, E. Möstl and E. Bamberg, *Anim. Reprod. Sci.* 5, 175 (1983).
[33] H. Kafali, M. Iriadam, I. Ozardali and N. Demir, *Arch. Med. Res.* 35, 103 (2004).
[34] N. Hirshfield, *Biol. Reprod.* 32, 979 (1985).
[35] S. G. Hillier, *Mol. Cell Endocrinol.* 179, 39 (2001).
[36] J. Duleba, R. Z. Spaczynski, D. L. Olive and H. R. Behrman, *Biol. Reprod.* 56, 891 (1997).
[37] J. Duleba, N. Foyouzi, M. Karaca, T. Pehlivan, J. Kwintkiewicz and H. R. Behrman, *Hum. Reprod.* 19, 1519 (2004).
[38] R. Farookhi, *Endocrinology.* 106, 1216 (1980).
[39] A. Bagnell, T. M. Mills, A. Costoff and V. B. Mahesh, *Biol. Reprod.* 27, 903 (1982).
[40] N. Isobe and Y. Yoshimura, *Theriogenology.* 53, 897 (2000).
[41] N. Isobe and Y. Yoshimura, *J. Reprod. Dev.* 53, 1119 (2007).
[42] Y. Ribadu, K. Nakada, M. Moriyoshi, W. C. Zhang, Y. Tanaka and T. Nakao, *Anim. Reprod. Sci.* 64, 21 (2000).
[43] J. B. Phogat, R. F. Smith and H. Dobson, *Anim. Reprod. Sci.* 55, 199 (1999).
[44] L. F. Archbald and W. W. Thatcher, *Large Dairy Herd Management,* Am. Dairy Sci. Assoc. Champaign (1992).
[45] J. Sirois and J. E. Fortune, *Endocrinology.* 127, 916 (1990).
[46] M. D. Calder, M. Manikkam, B. E. Salfen, R. S. Youngquist, D. B. Lubahn, W. R. Lamberson and H. A. Garverick, *Biol. Reprod.* 65, 471 (2001).
[47] N. R. Salvetti, L. A. Muller, J. C. Acosta, J. E. Gimeno and H. H. Ortega, *Vet. Pathol.* 44, 373 (2007).
[48] L. Johnson, *Anim. Reprod. Sci.* 78, 185 (2003).
[49] H. H. Ortega, M. M. Palomar, J. C. Acosta, N. R. Salvetti, B. E. Dallard, J. A. Lorente, C. G. Barbeito and E. J. Gimeno, *Res. Vet. Sci.* 84, 419 (2008).
[50] F. Rey, F. M. Rodriguez, N. R. Salvetti, M. M. Palomar, C. G. Barbeito, N. S. Alfaro and H. H. Ortega, *J. Comp. Pathol.* 142, 193 (2010).
[51] J. Hsueh, H. Billig and A. Tsafriri, *Endocr. Rev.* 15, 707 (1994).
[52] S. Y. Hsu and A. J. Hsueh, *Physiol. Rev.* 80, 593 (2000).
[53] A. Amsterdam, R. Sasson, I. Keren-Tal, D. Aharoni, A. Dantes, E. Rimon, A. Land, T. Cohen, Y. Dor and L. Hirsh, *Biochem. Pharmacol.* 66, 1355 (2003).
[54] M. R. Hussein, *Hum. Reprod. Update.* 11, 162 (2005).

[55] K. A. Slot, M. Voorendt, M. De Boer-Brouwer, H. H. Van Vugt and K. J. Teerds, *J. Endocrinol.* 188, 179 (2006).
[56] J. L. Tilly, *Rev. Reprod.* 1, 162 (1996).
[57] L. Johnson and J.T. Bridgham, *Reproduction.* 124, 19 (2002).
[58] E. Gürsoy, K. Ergin, H. Başaloğlu, Y. Koca and K. Seyrek, *Res. Vet. Sci.* 84, 56 (2008).
[59] E. Anderson and G. Y. Lee, *Tissue and Cell.* 29, 171 (1997).
[60] G. Almahbobi, C. Anderiesz, P. Hutchinson, J. R. McFarlane, C. Wood and A. O. Trounson, *Clin. Endocrinol.* 44, 571 (1996).
[61] K. Yacobi, A. Wojtowicz, A. Tsafriri and A. Gross, *Endocrinology.* 145, 1943 (2004).
[62] R. Robles, X. J. Tao, A. M. Trbovich, D. V. Maravel, R. Nahum, G. I. Perez, K. I. Tilly and J. L. Tilly, Endocrinology. 140, 2641 (1999).
[63] K. Yacobi, A. Tsafriri and A. Gross, *Endocrinology.* 148, 1717 (2007).
[64] M. D. Calder, B. E. Salfen, B. Bao, R. S. Youngquist and H. A. Garverick, *J. Anim. Sc.* 77, 3037 (1999).
[65] L. Pellatt, L. Hanna, M. Brincat, R. Galea, H. Brain, S. Whitehead and H. Mason, *J. Clin. Endocrinol. Metab.* 92, 240 (2007).
[66] A. Porter, S. L. Vickers, R. G. Cowan, S. C. Huber and S. M. Quirk, *Biol. Reprod.* 62, 62 (2000).
[67] A. Porter, R. M Harman, R.G. Cowan and S.M. Quirk, *Mol. Cell Endocrinol.* 176, 13 (2001).
[68] S. M. Quirk, R. G. Cowan, R. M. Harman, C. L. Hu and D. A. Porter, *J. Anim Sci.* 82, E-Suppl, E40 (2004).
[69] S. L. Vickers, R. G. Cowan, R. M. Harman, D. A. Porter and S. M. Quirk, *Biol. Reprod.* 62, 54 (2000).
[70] S. M. Quirk, R. M. Harman and R. G. Cowan, *Biol. Reprod.* 63, 1278 (2000).

In: Cell Proliferation
Editors: C. Zhang and X. Zeng
ISBN: 978-1-62417-352-3

Chapter V

Extracellular Protein-Induced Plant Cell Proliferation

***Anis Ben-Amar*[*1,2] *and Goetz M. Reustle*[2]**

[1]Laboratory of Plant Molecular Physiology, Center of Biotechnology, Borj Cedria Science and Technology Park, Hammam-Lif, Tunisia
[2]ALPLANTA - Institute for Plant Research, Neustadt/Weinstrasse, Germany

Abstract

Plant cell division occurs mainly in developing tissues and appears to be highly regulated. Factors affecting the core cell cycle have been extensively reviewed in recent years through cell culture systems.

Among several cell derived-substances and growth regulators, extracellular proteins commonly related to the cell surface matrix were widely involved in cell proliferation as well as in various aspects of plant growth and development. These proteins including arabinogalactan-proteins (AGPs), chitinases and pathogenesis-related proteins (PR proteins) were identified in a number of plant species within the micro-environment in which zygotic tissues and proliferating somatic embryogenic stem cells develop. Questions are raised about how these extracellular proteins control cell proliferation. Their expression and function are described highlighting cellular and physiological responses that are implicated in the regulation pathway of cell division. Molecular and biochemical insights are presented using a variety of model systems that have been already investigated.

The purpose of this review is to discuss the experimental results obtained to date to provide an overall outlook towards a better understanding of extracellular protein promotion of cell proliferation and totipotency in higher plants.

[*] E-mail address: anisbiotech@yahoo.fr.

1. Introduction

Plant growth and development depends essentially on the stem cell system found in meristematic tissues whose activity consist to produce new cells giving to plants their exceptional feature of indefinite growth. Thus, apical meristems contain a pool of pluripotent stem cells and an amplifying compartment of proliferating cells [1, 2]. In contrast to animals, organogenesis in plants is a postembryonic and continuous process. Single differentiated plant cells can revert to a juvenile state; proliferate and trans-differentiate again. This exceptional potential illustrates the ability of certain cells to produce undifferentiated clusters or fully totipotent somatic embryos that could regenerate mature plants. Although the overall cell cycle is highly conserved, multicellular organisms have imposed extra layers of complexity based on the balance between cell proliferation and differentiation: two distinct morphogenetic pathways [1, 3].

In pluripotency, a single cell gives rise to most, but not all, of the various cell types that make up a plant. In totipotency, a single cell can develop into an embryo (under certain conditions), thereby producing a new adult organism. During both of theses programs, a single differentiated somatic cell re-enters the cell cycle *via* the cell-reactivation process [4]. Once induced, these competent cells under appropriate *in vitro* culture conditions become committed to different morphogenetic pathways such as organogenesis (pluripotency) or somatic embryogenesis (totipotency).

Cell proliferation relies on mitosis known also as the cell division cycle, the process that enables a cell to produce two daughter cells. Cell cycle progression is characterized by a series of unidirectional events that drive a cell from its birth to its division [3]. The cellular environment is extremely important for its role in signaling process. Based on this fact, culture of isolated plant cells led to their separation from almost all morphogenetic interactions. This state is accompanied by acquisition of new cell fate that requires an extensive reprogramming of gene expression [5].

This pathway depends in some manner on the dynamic of extracellular matrix involved in cell signaling and cell-to-cell interactions that could modulate morphogenesis. In fact, plant cells secrete several compounds that form the extracellular matrix (ECM). This matrix is an elaborate covering outside the plasma membrane which consists in a sticky layer of embedded glycoproteins. It is a meshwork-like substance found within the extracellular space and in association with the cell surface. Beside its role in maintaining cell shape and providing a supporting structure for cell architectural design, it is mainly implicated in a number of aspects including cell adhesion, recognition, signaling, developmental process and control of plant morphogenesis (Figure 1).

Extensins, glycine-rich proteins (GRPs), proline-rich proteins (PRPs), lectins, arabino-galactan proteins (AGPs) with chitinase/glucanase pathogen-related (PR) proteins are the most abundant and to date the widely documented plant cell wall proteins [6, 7]. During the last two decades, a number of research groups in cell wall matrix - including Showalter (Ohio, USA), Nothnagel (CA, USA), Haseloff (Cambrige, UK), Roberts (Norwich, UK), Knox (Leeds, UK), De Vries (Netherlands), Bacic (Australia) and Seifert's group (Austria) - have been contributing substantially to rapidly advancing knowledge of plant cell wall biology. Cell wall proteins are cross-linked into the wall and have structural functions [8] as well as they are concerned by signaling and interactions with plasma membrane proteins [9].

The expression of these extracellular proteins (EPs) at the cell surface is tissue/organ specific and can exert a regulatory effect over cell behavior [10]. Each tissue is characterized by his own set of cell wall proteins related to his developmental status and controlled by morphogenetic signaling since the cell wall provides countless potential sources of information [7]. ECM proteins play crucial roles in plant development, morphogenesis, cell division and proliferation [11]. Evidence is increasing about implication of extracellular proteins in the induction of cell cycle machinery. In this chapter, we investigate the mechanism underling the control of cell proliferation in cultured plant cells in order to monitor the involvement of extracellular matrix components in such processes.

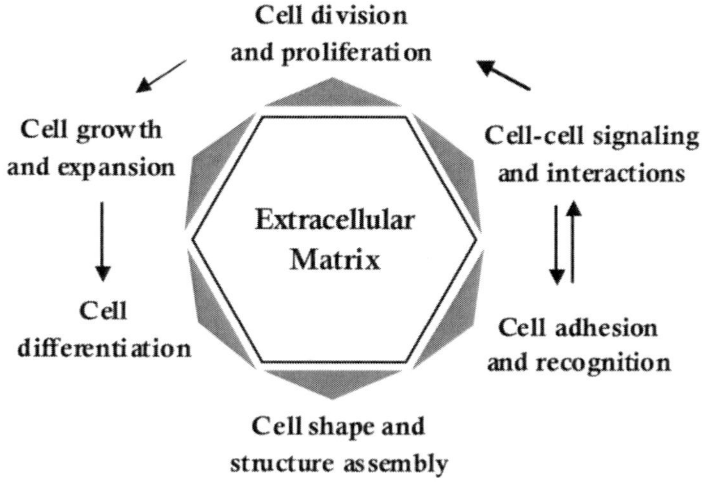

Figure 1. Biological activities modulated by the extracellular matrix of plant cells.

2. Cell Systems to Investigate Cell Proliferation

Cell division could be studied in plant tissue materials in a proliferative state such as meristematic and zygotic tissues, in undifferentiated cell clusters, protoplast suspensions and embryogenic cells. Tumor cell proliferation mediated by *Agrobacterium tumefaciens* and *rhizogenes* causes the development of a crown gall structure that could be used to better understand mechanism underlying the cell cycle. In a similar way, *Rhizobium* nodule and arbuscular mycorhizal infection site-derived symbiotic association are also used as target candidates to investigate how extracellular proteins might influence cell proliferation. In both systems, stem cells and cancer or symbiotic cells share an ability to divide continuously without undergoing senescence [12]. In such experimental systems, the environment is controllable and the population of target cells should be as homogeneous as possible. Since these requirements are difficult to meet when using intact plants as experimental systems, *in vitro* plant cell cultures might be an alternative to meet such requirements [13]. Within available tissue culture systems, embryogenic cells were shown to be the most suitable target material for plant genetic manipulation and gene transfer technology. In fact, plant somatic cells have the ability to undergo sustained divisions and give rise to an entire organism. This remarkable feature is called plant cell totipotency. Somatic embryo is a notable illustration of plant totipotency and involves reprogramming of development in somatic cells toward the embryogenic pathway [14]. Hence, embryogenic cell suspensions appear to provide a valuable experimental tool for studying how cells acquire totipotency.

3. Conditioned Medium Enhances Cell Division during Somatic Embryogenesis

Embryogenic cell suspensions represent the best way to study extracellular proteins and their effect on cell division and further development. Changes in these protein expression patterns have been associated with induction of somatic embryogenesis [15, 16]. Besides growth regulators known to induce embryo formation, other classes of proteins have been identified as embryo-stimulating factors, especially those secreted in the culture medium [17]. In fact, this conditioned medium, in which cells have been grown previously, contain a mixture of extracellular proteins that play a significant role in the establishment and maintenance of highly proliferating embryogenic cell lines [18]. The relationship between extracellular protein pattern and cellular proliferation has been recently demonstrated using plant cell culture systems [18, 19].

Using conditioned medium, synchronized cell-division systems have been established in suspension cultures of grapevine [18]. Furthermore, in bioassays concentrated extracellular protein fractions derived from growth medium of cultured grapevine cells were supplemented to fresh medium and resulted in a promoting effect on suspension initiation and cell proliferation. Interestingly, Ben-Amar and coworkers [18] reported also that conditioned medium derived from one cell line has been shown to be effective on other lines in an intra- and inter-specific manner suggesting that extracellular proteins are likely genotype-independent. Somatic embryogenesis was significantly induced and different embryo developmental stages (globular, heart and cotyledon) could be identified in suspension cultures. Further examination revealed that extracellular proteins present either in the culture medium or in the cell extracts efficiently induced proliferation of low-growing cell lines [19]. Taken together, these results confirmed again the occurrence of extracellular proteins on the cell surface that could be secreted into the medium and play an important role to promote cell division.

4. Chitinases/PR Proteins Involvement in Cell Proliferation

Several reports indicated that proteins secreted into the medium accompany the formation of embryogenic cell clusters. So far, three different extracellular proteins (EP_1, EP_2 and EP_3) have already been identified in carrot cell cultures. Van Engelen et al. [20] firstly described an extracellular protein EP_1 that is only secreted by non-embryogenic cells. Sterk et al. [21] reported another extracellular protein EP_2 of 10-kDa, identified as a lipid transfer protein (LTP) and produced by embryogenic cells. Lipid transfer proteins (LTPs) are largely secreted extracellularly during somatic embryogenesis of grape and carrot cells [15, 22] and may have a function in transporting phospholipids to various cellular locations. These pathogen-related proteins are believed to be involved in plant defense mechanisms with upregulation of its expression in response to fungal elicitors [23]. *In situ* hybridization also demonstrates the localization of EP2 expression to the protodermal layer of somatic and zygotic embryos; thus EP2 may serve as an early marker for embryogenesis [21]. Later, an EP_3 of 32-kDa has been purified and identified as a glycosylated class IV endochitinase based in its ability to rescue proliferation and to lift the arrest of somatic embryos of mutant carrot cell line ts11 at non-permissive temperatures [24, 25].

Chitinases represent one of the main classes of pathogen-related protein (PR proteins) family in plants that have been reported during plant-microbe interactions in case of leguminous-*Rhizobium* symbiosis [26] or through association with arbuscular mycorrhizal fungi [27]. Chitinases are shown to generate signal molecules acting as messengers modulating cell cycle activity and developmental programs. Their action is suggested to be involved in cell proliferation [25, 28]. Exogenous addition of purified chitinase resulted in an increase of cell populations and multiplication rate up to five-fold in suspension of proembryogenic masses [18, 19]. The growth promoting effect is mainly attributed to the cleavage of oligosaccharide residues playing a key role as signal molecules [29, 30, 31]. The lipochito-oligosaccharides thus generated were identified as Nod-factors, by which chitinases control the biological activity of undifferentiated cells and induce infection-related early nodulin gene expression in pea root hairs [32]. In fact, it was proposed that a 32-kDa endochitinase identified from carrot cell cultures could degrade specific precursor molecules having N-acetylated glucosamine residues present in a low amount in the plant cell wall [24, 26]. Thus, glycoproteins containing N-acetylglucosamine were identified as good candidates to serve as substrates for chitinase [33].

Moreover, chitinases were reported to induce a 38-kD extracellular SER (somatic embryogenesis-related) protein already characterized in chicory [34]. This polypeptide coding for ß-1,3-glucanase was expressed at higher level in the medium of embryogenic suspensions than in the non-embryogenic lines [35]. Such specific extracellular protein release is essential during plant embryogenesis pathway and confirms the implication of chitinases in the preservation of an embryogenic state. This also indicates that extracellular proteins encoded by such embryo-specific genes may have a potential role in callose degradation to maintain cell wall plasticity which ultimately promotes cell division and proliferation.

5. AGP-mediated Promotion of Cell Proliferation

Among extracellular proteins, glycoproteins and especially arabinogalagtan proteins (AGPs) are well represented and undoubtedly one of the most complex families of macromolecules found in plants [36]. AGPs are highly glycosylated proteoglycans (often more than 90% carbohydrate) that belong to the hydroxyproline-rich glycoproteins. They are commonly distributed in the plant kingdom and their complexity arises from the incredible diversity of the glycans decorating the protein backbone [37]. At subcellular level, AGPs were found in the plasma membrane [38], in the cell wall [39] and into the medium of cultured cells [18, 19]. The widespread distribution of AGPs at the plant cell surface implicates their important role in plant development and morphogenesis [40]. There are contradictory data about their exact functions and possible involvement in a broad variety of processes as plant growth and development, plant stress response [41, 42], signal transduction and plant-microbe interactions [43], cell proliferation [19, 39, 44, 45], cell expansion and differentiation [29, 46, 47, 48], cell death [40] and somatic embryogenesis [18, 30, 49].

Since two decades, AGPs have attracted extensive attention from plant biologists following reports of their involvement in plant development. Their structure, expression and potential roles as key regulators at the cell surface were summarized in several reviews [36, 37, 50, 51]. Moreover, AGPs cross-linked to other cell wall components have an important role in cell communication and signaling network within the extracellular interface. Gaspar et al. [43] proved that reduced expression of lysin-rich AGPs resulted in a decreased *Agrobacterium*-mediated transformation efficiency suggesting the implication of AGPs in binding *Agrobacterium* at the cell surface allowing thus a higher transformation. Level of extracellular AGPs was measured spectrophotometrically using ß-D-glucosyl Yariv's reagent which selectively binds to AGPs and produces a complex that precipitates at room temperature [62]. Therefore, addition of Yariv's reagent blocks the biological activity of AGPs and dramatically inhibited cell division as well as somatic embryogenesis in a concentration-dependent manner probably due to the arrest of cell cycle progression [4, 18, 39].

In fact, in cell culture systems, evidence of a possible implication of AGPs in the proliferation of *Rosa* cell cultures was firstly proposed by Serpe and Nothnagel [39] and may be related to other roles of AGPs in embryogenesis and morphogenesis which entail regulation of cell division. Later, similar studies on carrot [29, 45], sugar beet [53], tobacco [54], grapevine [18], zucchini squash [19] and chicory [4] described the involvement of AGPs in cell proliferation. In fact, our previous results indicated that conditioned medium containing AGPs were associated with maintaining high proliferating cell suspensions [18, 19]. Conversely, inactivation of soluble AGPs or reduction of their availability in the medium alters the progress of cell division in undifferentiated cell cultures by an arrest of the cell cycle [55]. Such inactivation could also induce programmed cell death in *Arabidopsis* cells within 3 days by higher concentrations of Yariv reagent [40, 50].

AGPs may also be involved in the establishment of embryogenic pattern. The embryogenic potential of old carrot cultures that had gradually lost their capacity can be restored by carrot seed extracts containing AGPs [49]. Next to that, purified AGPs from carrot embryogenic cultures even at nanomolar concentrations can reinduce non-embryogenic masses to embryo-forming cells [34]. It was found that cell cycle was rescued by AGP application when other investigators proved that it was possible to manipulate somatic embryogenesis by addition of exogenous AGP-like substances [18]. Similarly, Lucau-Danila et al. [4] showed that Yariv reagent was able to block somatic embryogenesis as the main totipotent cell-producing process. The embryos appeared very rapidly following transfer to medium lacking Yariv which indicate that such inhibition of cell cycle process is reversible.

Wisniewska and Majewska-Sawka [56] reported that AGP-rich extracts isolated from medium of cultured sugar beet cells are able to enhance the organogenesis of guard protoplast-derived callus. Similarly, the addition of zucchini squash cell extract derived from fast-growing cell cultures could enhance multiplication rate of slow-growing cell lines. Our previous work demonstrated that the responsible factor of this effect was identified as AGPs by immunoblotting with MAC 207 anti-AGP monoclonal antibody [19]. In particular, monoclonal antibodies directed against carbohydrate epitopes of AGPs gave evidence about stage- and tissue-specificity by immune-detection and subcellular localization techniques [46, 53]. These data may be further used for predicting emerging tissue patterns or developmental fate of cells [45].

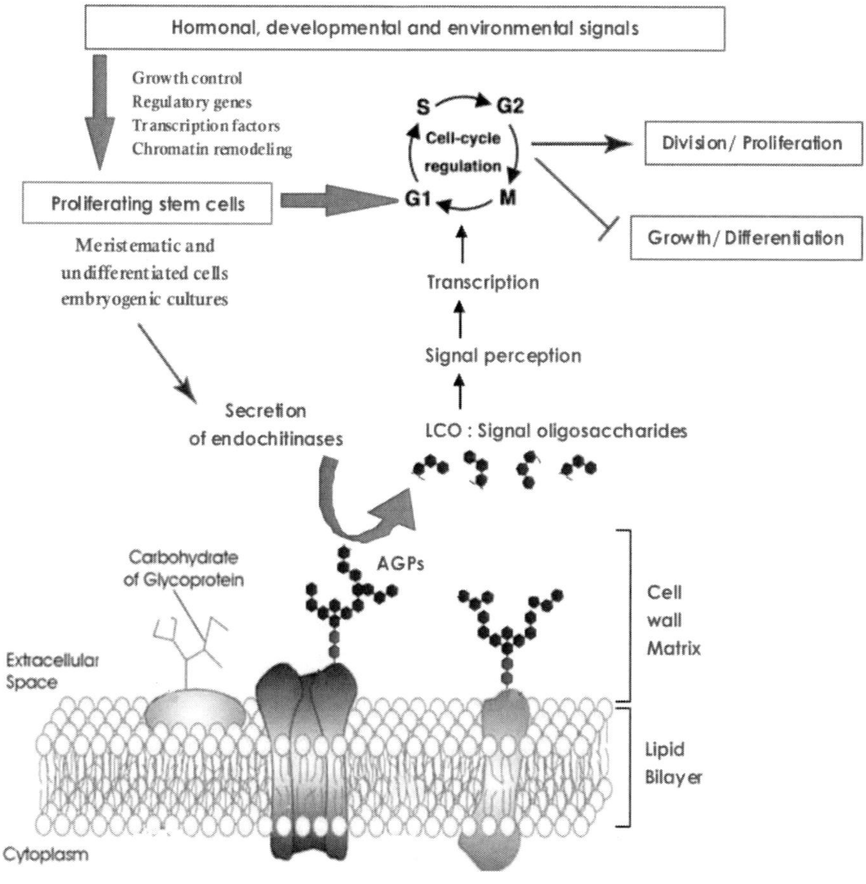

Figure 2. Schematic view dissecting plant cell proliferation network in relation with the extracellular matrix. The morphogenetic pathway depends on internal (hormonal, developmental) and external (environmental) signals, which in many cases target components of the cell cycle regulatory machinery [3]. A proposed model of AGPs-mediated cell proliferation is illustrated. This model proposes that endochitinases secreted by plant cells are activated to split carbohydrate side chains of AGPs in small molecules called lipochito-oligosaccharides (LCOs) which could act as signaling messengers to promote cell division [17, 18, 19]. Thus proliferation pathway is turned on while cell growth, expansion and differentiation are switched off with acquisition of pluripotentiality driven by chromatin remodeling to induce new cell fates.

AGPs have been also reported as substrate of cell wall enzymes to produce some components with hormone-like activity which may contribute in developmental regulation and cell division (see for review [36]). Van Hengel et al. [57] claimed for an AGP splitting by endochitinases that generate lipochito-oligosaccharides (LCO)-like molecules homologous with *Rhizobium*-Nod factors [26, 58] implicated in signaling pathway and controlling proliferation of plant cells. This finding was described in similar studies (see for review [17]) and later confirmed by Ben-Amar and coworkers [18, 19] respectively in grapevine and zucchini squash suspension cultures. The figure 2 summarizes the action of extracellular proteins and especially the chitinase-mediated AGP degradation to generate LCO molecules stimulating cell proliferation in hormonal-like manner.

AGPs might act as paracrine and autocrine signals in many biological processes (for review, see [37, 51]). Their potential to bind to β-glycan polymers together with their plasma membrane localization put them in a strategic position to mediate between cell wall polymers and cell signaling. Different signals lead to the cleavage of the anchor and release of soluble AGP monomers through the cell wall into the growth medium in suspension cultures or into the middle lamella and the intercellular space of cells. The continuous release of AGPs is characteristic for rapid growing plant cells. Their signaling potential as well as their biological activity in nanomolar concentration let us believe about their relationship or their connection with plant growth regulators. It is tempting to speculate their potent action in a crossroad of developmental and cellular network. Extracellular AGPs appeared to be a broad proteoglycan family of molecules with highly regulated expression which is an indication for their significance in plant development.

6. Other Extracellular Proteins Identified during Proliferation

Among ECM proteins identified through cellular proliferation, several nodule-enhanced genes or nodulins encoding repetitive proline-rich cell wall proteins were induced during early interactions with rhizobia [59]. Nodulin- and putative extensin-mediated cell proliferation was strongly correlated with a massive restructuration of the ECM associated to infection and nodulation [60, 61]. Additionally, an extracellular EXO protein was reported as a potential mediator of brassinosteroid-promoted growth and mediated cell expansion in *Arabidopsis* leaves [62]. Tian et al. [11] identified further proteins based on differential proteomic analysis of soluble extracellular proteins and specific identification by mass spectrometry: a cystatin and cystein protease, both involved in proliferation of cell suspension in rice.

Receptor-protein kinases (RPKs) allowing cells to recognize and respond to their extra-cellular environment have been also raised as signaling proteins containing extracellular leucine-rich-repeat (LRR) domains, and widely associated to embryogenic competence and cell division [63]. Matsubayashi et al. [64] reported an LRR-receptor kinase involved in the perception of peptide plant hormone, phytosulfokine inducing cell expansion and proliferation [65]. Furthermore, wall-associated kinases (WAKs) connecting the cytoplasm to the ECM, possess extracellular domains that include epidermal-growth factor (EGF)-like repeats and extensin motifs. Somatic embryogenesis-receptor kinases (SERKs) have been also identified in a differential screen to identify genes specifically expressed in embryogenic cell cultures of carrot [66], *Arabidopsis* [67] and grapevine [68] among many plant species. It was also shown that SERK protein is transiently expressed during embryogenesis in proembryogenic masses and disappearing after the globular stage, indicating that SERK is required to make cells competent to undergo embryogenesis and that this competence reflects the ability to perceive an inductive signal [63].

7. Cell Proliferation versus Cell Differentiation

To properly understand the mechanism of plant morphogenesis, it is necessary to have biological systems that allow simple and direct observation of developmental processes. Most previous studies have adopted a simplified plant system to investigate how cell proliferation and expansion is driven by switching on and off correspondent genes or transcription factors [69, 70]. Cell division and differentiation are highly coordinated during morphogenesis [1]. Morphogenesis in plants arises from the interplay of genetic and physical interactions within a growing network of cells. The physical aspects of cell proliferation and differentiation are genetically regulated, but constrained by mechanical interactions between the cells which implicate an elaborate three-dimensional extracellular matrix [70].

In recent years much attention has been paid to plant cell culture as tool to study cell division. How plant cell proliferation is regulated within the context of development has been a subject of debate for many years. Growth of plant organs relies on coordinated cell proliferation followed by cell growth, but the nature of the cell-to-cell signal that specifies organ size remains elusive [71]. Cytoplasmic cell growth and cell division in plant cells are regulated independently but they are coupled so that growth is required for normal proliferation to produce daughter cells with fixed sizes [72]. Over the past decade, mutants have been found in which either cell division or cell expansion is perturbed. Molecular studies strongly support the notion that a compensatory mechanism exist whereby cell expansion can be enhanced to cover defective cell division [73]. Plants cannot develop in the absence of cell division; as a result aberrant cell-cycle mutations cause embryo lethality [74].

It is well established that most differentiated plant cells do not lose their potentialities during normal development but retain plasticity and are still able to dedifferentiate towards acquiring new fates [5]. The remarkable regenerative property displayed by plants is largely based on the capacity of somatic cells to undergo regeneration from differentiated tissues. Most studies on the dedifferentiation process have been carried out in a protoplast system. Acquisition of pluripotency occurs during protoplast formation and is characterized at subcellular level by extensive chromatin decondensation and histone modifications [5]. The nucleus volume is larger than that of a differentiated cell as a result of the relaxed chromatin structure associated with transcriptional activity [75].

Proliferating cells often switch within a period of time to the differentiation pathway. A quite unique situation occurs in a suspension culture of dedifferentiated proliferating cells that can be maintained for years while the cells remain in an undifferentiated state as for embryogenic cell suspensions [13, 18]. In this case, cell cycle-specific transcriptional programs involve a switch on and off of a set of genes required to progress from one stage to the next one [75]. In cell cultures subjected to stress, secretion of ethylene inhibits cell cycle progression and when the stress persists, cells exit the mitotic cell cycle and initiate the differentiation process [76]. These data present a conceptual framework to understand how environmental stress reduces plant growth.

While extracellular proteins released into the medium of embryogenic cultures highly promote cell proliferation, accumulation of these proteins was found to completely blocks embryo differentiation [15, 18]. In grapevine embryogenic suspension cultures, change of the medium everyday reduce the amount of extracellular proteins and as a result enhance regeneration process [18]. It has been shown that a 10-kDa LTP-like protein was largely accumulated until 30% of total extracellular proteins in grape and carrot embryogenic suspensions in auxin-free medium and arrested embryo development at heart stage [15, 21]. Thus, we suggested that extracellular proteins inducing intensive proliferation became probably in different environmental and hormonal conditions inhibitors of embryo differentiation and subsequent growth [18]. These findings clearly indicate that cell proliferation and differentiation appear as two distinct morphogenetic pathways which require different environmental conditions of secreted proteins and metabolites.

8. Other Related Aspects of Cell Proliferation Control

Growth Regulators Coordinate Proliferation of Plant Cells

Hormonal signaling in cell-matrix interface determines the timing of cellular survival, proliferation, differentiation and apoptosis. We do not know why cells in close proximity behave differently. One possible explanation is that they differ in their hormonal content. Auxin/cytokinin balance was often recognized to influence cell division and proliferation in plant tissue cultures and already known to be determinant of morphogenesis controlling developmental changes in plant cells. A synchronized cell-division system requires a correct combination of such regulators, such as auxins and cytokinins frequently considered as they regulate the cell cycle and trigger cell divisions [77]. Auxin is likely a central factor that coordinates multiple aspects of growth and cell division [72]. Differential auxin distribution has been proposed to act as a morphogen to set up distinct zones for cell division, cell expansion and differentiation.

We previously observed a positive feedback of extracellular protein derived from conditioned medium and stimulating proliferation of cell suspensions cultured in auxin-containing media [18, 19]. Obviously, high auxin concentrations promote cell division in cell culture systems [78]. Contrary to this, in auxin-free medium, cells accumulate in the G_1 phase and cell division is blocked. Addition of auxin to the medium releases the G_1 arrest and stimulates progression of the cell cycle as cells could divide in synchrony [13]. Influence of auxin on extracellular protein pattern in suspension cultures is considered [15, 16]. In somatic embryogenesis, 2,4-D a strong synthetic auxin is often required for the initiation and maintenance of proembryogenic masses in continuous proliferating state [15, 18]. The removal of 2,4-D switches cells from proliferation to elongation which is known as an aspect of differentiation [29].

Cytokinins are also growth substances that promote cytokinesis and plant cell division. Several lines of evidence indicate their requirement for cell proliferation *in vitro* and *in vivo* via increasing the plasticity of the cell wall [79, 80, 81]. Cytokinins were discovered in relation to their ability to stimulate cell division in tissues supplied with an optimal level of auxin. Evidence suggests that both auxin and cytokinins participate in regulation of the cell cycle and that they do so by controlling the activity of cyclin-dependent kinases (see section below 8.3). Next to these two common growth regulators (auxins and cytokinins), some substances with hormone-like activity that might interact with cell wall proteins and promote cell division have been frequently reported such as phytosulfokines and brassinosteroids among others [64, 65, 82, 83]. Currently, phytosulfokines were shown to be associated with proliferation, growth and expansion of plant cells as their disruption affects cellular longevity and potential of growth [84, 85, 86]. Recent studies illustrate also that brassinosteroids play crucial role in multiple process of plant growth, cell division, expansion and signaling implicating plant ECM [82, 87].

Cryopreservation Enhances Cell Proliferation

Beside hormonal control, the stress caused by cryopreservation could induce cell proliferation via indirect reactivation of the cell cycle. Ongoing studies revealed that embryogenic cells subjected to dehydration/encapsulation treatment during cryopreservation showed a high proliferation potential comparatively to control cells (A. Ben-Amar, unpublished results). Cryopreservation was used to circumvent the loss of embryogenic competence and to reinduce cell totipotency *in vitro* [88]. It has been also reported that embryogenic potential was largely improved in cryopreserved plant cells showing an intense morphogenetic competence and an increase in cell proliferation in banana [89], grapevine [90] and *Gentiana* [91]. As a possible explanation, once dehydration is considered as an osmotic stress, this desiccation could probably up-regulate chitinase and/or glucanase expression and release of AGPs in the culture medium. This finding was reported in tobacco Bright Yellow-2 cells subjected to salt stress [42] supporting the prediction that AGPs might modulate cell proliferation in relation with plant stress response. By maintaining intact the main structural elements of the extracellular matrix and membrane integrity, cryopreservation was shown to increase the proliferation rate of human [92] and plant cells (A. Ben-Amar, unpublished data) via a mechanism involving ECM components.

Molecular Regulation of Cell Proliferation and Differentiation

Cell growth and proliferation are part of a regulatory cascade that is controlled by developmental genes, Homeobox sequences, firstly discovered by McGinnis et al. [93] and known as the key regulatory genes controlling morphogenetic pattern in multicellular organisms. Since, MADS-box homeotic genes have been well-characterized in plants such as *Arabidopsis*, maize and rice [94]. The encoded homeodomain coding transcription-factors with DNA-binding activities is associated with regulation of morphogenetic processes including cell proliferation and differentiation [34]. In this context, the homeobox VOX_{11} gene, found to be an auxin- and cytokinin-responsive gene, is required to activate crown root development. This gene was expressed in emerging crown roots and in cell division regions such as meristematic tissues [95]. Loss-of-function mutation or down-regulation of the VOX_{11} gene reduced the number and the growth rate of crown roots, whereas its over-expression was reported to induce precocious crown root growth and dramatically increased the root biomass by enhancing cell division. Furthermore, membrane protein kinases were found to promote proliferation of *Arabidopsis* procambial cells and suppress their commitment to xylem differentiation via activation of VOX_4-related homeobox gene [96].

Progression through the cell cycle is driven by conserved kinases designed as cyclin proteins [72]. They appeared to be involved in the G_1/S transition, regulation of S phase and control of entry into mitosis [97]. Each unidirectional transition is controlled by a subset of cyclin-dependent protein kinases: CDKs [73]. Cyclins regulate cell division by increasing endogenous level of cytokinins suggesting an inter-dependent relationship between cyclins and cytokinins [80]. Accordingly, overexpression of a CDK-inhibitor protein blocks the cell cycle but not growth leading to enlarged cells [98], similarly to what was observed when CDK proteins rapidly disappear since cells are transferred to hormone-free medium [99]. Recent work has shown that ethylene affects cell cycle progression via inhibition of CDK-A [76]. It has been also demonstrated that elevated expression level of E_2FB transcription factor was able to activate cell division resulting in overproliferation of both *Arabidopsis* and BY-2 tobacco cells respectively [69, 72].

Although these efforts at the molecular level, little is known about how developmental programs control cell division. Rodriguez et al. [100] found that increase in the amount of a 21nt small interfering RNA species (miR396) attenuate cell proliferation in mature leaves of *Arabidopsis* through the repression of growth regulating factors. Conversely, *Arabidopsis* plants over-expressing the *Agrobacterium* Ti plasmid 6b proteins display micro-RNA deficiency and stimulate therefore plant cell division associated with histone hypomethylation in undifferentiated cells [101]. These observations revealed the implication of RNA interference, DNA methylation and histone modifications in reprogramming the epigenome of proliferating cultured cells. In opposition, other experiments carried out on animal cells in which components essential for miRNA biogenesis were disrupted in flies suggest that miRNAs may play a role in stem cell maintenance. Specifically, dcr-1-deficient GSCs inappropriately activated the G1/S cell cycle checkpoint, suggesting that miRNAs normally act to promote continuous cell cycle progression [12]. Despite its relevance for studying the molecular basis of cell proliferation, the fundamental mechanism underling cell cycle activation in proliferating cells remains to date poorly understood.

Conclusion and Future Outlooks

Extracellular matrix plays a major role in cell proliferation and morphogenesis during plant development. Besides hormonal control, several secreted proteins described above and connected to the cell wall seem to be involved in cell division machinery. Complex interaction between these cell wall glycoproteins and their sub-products are required for correct cell signaling pathways. Despite the progress achieved during the last years and focusing to explore the function of cell wall components and their possible regulation of cell cycle progression and cell fate, the overall network is still unknown. However, several recent studies strongly indicate their direct or indirect involvement in cell proliferation.

Molecular dissection of plant morphogenesis is providing new insights into the possible mechanisms that could regulate cell proliferation. The challenge now is to investigate more the biological function of these extracellular matrix proteins strongly associated to cell competence, pluripotency and cell division pathway. This is being attempted either by creating transgenic plants that express an antisense construct or by working with genes that have already been disrupted through loss-of-function mutations.

Acknowledgments

We thank Prof. J. Paul Knox (Center of Plant Sciences, University of Leeds, UK) for critical reading the manuscript and Prof. Sacco deVries (Pant Cell Biochemistry, WU Agrotechnology & Food Science, Wageningen, Netherlands) for his suggestions and comments.

References

[1] J. Fleming, *J. Exp. Bot. 57*, 25 (2006).

[2] Scheres, *Nature Rev. Mol. Cell Biol. 8*, 345 (2007).

[3] Gutierrez, *Nature Cell Biol.* 7, 535 (2005).

[4] Lucau-Danila, L. Laborde, S. Legrand, L. Huot, D. Hot, Y. Lemoine, J. L. Hilbert, S. Hawkins, M. C. Quillet, T. Hendriks, A. S. Blervacq, *BMC Plant Biol. 10*, 122. doi: 10.1186/1471-2229-10-122 (2010).

[5] G. Grafi, *Developmental Biol. 268*, 1 (2004).

[6] Showalter, *Plant Cell 5*, 9 (1993).

[7] G. J. Seifert and C. Blaukopf, *Plant Physiol. 153*, 467 (2010).

[8] G. I. Cassab, *Ann. Rev. Plant Physiol. 49*, 281 (1998).

[9] Jamet, H. Canut, G. Boudart, and R. F. Pont-Lezica, *Trends Plant Sci. 11*, 33 (2005).

[10] K. Roberts, *Curr. Opin. Cell Biol.* 6, 688 (1994).

[11] L. Tian, L. Zhang, J. Zhang, Y. Song and Y. Guo, *Biochemica and Biophysica Acta 1794*, 459 (2009).

[12] H. W. Hwang and J. T. Mendell, *Brit. J. Cancer 94*, 776 (2006).

[13] H. Fukuda, M. Ito, M. Sugiyama and A. Komamine, *Int. J. Dev. Biol. 38*, 287 (1994).

[14] O. Karami, B. Aghavaisi and A. M. Pour, *J. Chem. Biol. 2*, 177 (2009).

[15] P. Coutos-Thevenot, O. Maes, T. Jouenne, M. C. Mauro, J. P. Jouanneau, M. Boulay, A. Deloire and J. Guern, *Plant Sci. 86*, 137 (1992).

[16] O. Maes, P. Coutos-Thevenot, T. Jouenne, M. Boulay and J. Guern, *Plant Cell Tiss. Org. Cult. 50*, 97 (1997).

[17] Matthys-Rochon, *Acta Biol. Cracov. 47*, 23 (2005).

[18] Ben-Amar, P. Cobanov, K. Boonrod, G. Krczal, S. Bouzid, A. Ghorbel and G. M. Reustle, *Plant Cell Rep. 26*, 1439 (2007).

[19] Ben-Amar, P. Cobanov, A. Ghorbel, A. Mliki and G. M. Reustle, *Biol. Plant. 54*, 321 (2010).

[20] A. Van Engelen, P. Sterk, H. Booij, J. H. G. Cordewener, W. Rook, A. Van Kammen and S. C. De Vries, *Plant Physiol. 96*, 705 (1991).

[21] P. Sterk, H. Booij, G. A. Schellekens, A. Van Kammen and S.C. De Vries, *Plant Cell 3*, 907 (1991).

[22] J. Fleming, T. Mandel, S. Hoffmann, P. Sterk, S. C. De Vries and C. Kuhlemeier *Plant J. 2*, 855 (1992).

[23] E. Gomès, E. Sagot, C. Gailard, L. Laquitaine, B. Poinssot, Y. H. Sanejouand, S. Delrot and P. Coutos-Thevenot Mol. *Plant Microbe Interact. 5*, 456 (2003).

[24] J. De Jong, J. Cordewener, F. Lo Schiavo, M. Terzi, J. Vandekerckhove, A. Van Kammen and S. C. De Vries, *Plant Cell 4*, 425 (1992).
[25] J. Van Hengel, F. Guzzo, A. B. Van Kammen, S. C. De Vries, *Plant Physiol. 117*, 43 (1998).
[26] J. De Jong, R. Heidstra, H.P. Spaink, M. V. Hartog, E. A. Meijer, T. Hendriks, F. Lo Schiavo, M. Terzi, T. Bisseling, A. B. Van Kammen and S. C. De Vries, *Plant Cell 5*: 615 (1993).
[27] M. L. Van Buuren, I. E. Maldonado-Mendoza, A. T. Trieu, L. A. Blaylock and M. J. Harrison, *Mol. Plant Microbe Interact. 12*, 171 (1999).
[28] M. I. Tchorbadjieva and I. Y. Pantchev, *Biol. Plant. 50*, 142 (2006).
[29] W. G. T. Willats, C. G. Steele-King, S. E. Marcus and J. P. Knox, *Plant J.* 20, 619 (1999).
[30] J. Van Hengel, Z. Tadesse, P. Immerzeel, H. Schols, A. Van Kammen and S. C. De Vries, *Plant Physiol. 125*, 1880 (2001).
[31] Kasprzewska, *Cell. Mol. Biol. Lett. 8*, 809 (2003).
[32] Horvath, R. Heidstra, M. Lados, M. Moerman, H. P. Spaink, J. C. Promé, A. Van Kammen and T. Bisseling, *Plant J. 4*, 727 (1993).
[33] R. Malinowski and M. Filipecki, *Cell. Mol. Biol. Lett. 7*, 1137 (2002).
[34] Chugh and P. Khurana, *Curr. Sci. 83*, 715 (2002).
[35] S. Helleboid, T. Hendriks, G. Bauw, D. Inze, J. Vasseur and J. L. Hilbert, *J. Exp. Bot.* 51, 1189 (2000).
[36] Showalter, *Cell. Mol. Life Sci. 58*, 1399 (2001).
[37] M. Ellis, J. Egelund, C. J. Schultz and A. Bacic, *Plant Physiol. 153*, 403 (2010).
[38] P. Komalavilas, J. K. Zhu and E. A. Nothnagel, *J. Biol. Chem. 266*, 15956 (1991).
[39] M. D. Serpe and E. A. Nothnagel, *Planta 193*, 542 (1994).
[40] M. Gao and A. M. Showalter, *Plant J. 19*, 321 (1999).
[41] Y. Guan and E. A. Nothnagel, *Plant Physiol. 135*, 1346 (2004).
[42] T. A. Lamport, M. J. Kieliszewski and A.M. Showalter, *New Phytol. 169*, 479 (2006).
[43] Y. M. Gaspar, J. Nam, C. J. Schultz, L. Y. Lee, P. R. Gilson, S. B. Gelvin and A. Bacic, *Plant Physiol. 135*, 2162 (2004).
[44] D.V. Basile and M. R. Basile *Mem Torrey Bot. Club 25*, 63 (1993).
[45] J. M. Thompson and J.P. Knox, *Planta 205*, 302 (1998).
[46] J. P. Knox, P. J. Linstead, J. Peart, C. Cooper and K. Roberts, *Plant J. 1*, 317 (1991).
[47] K. J. D. Lee, Y. Sakata, S. L. Mau, F. Pettolino, A. Bacic, R. S. Quatrano, C. D. Knight, and J. P. Knox Plant Cell *17*, 3051 (2005).
[48] W. G. T. Willats and J. P. Knox, *Plant J. 9*, 919 (1996).
[49] M. Kreuger and G. J. Van Holst, *Planta 197*, 135 (1995).
[50] Majeska-Sawka and E. Nothnagel, *Plant Physiol. 122*, 3 (2000).
[51] Seifert, K. Roberts, *Annu. Rev. Plant Biol. 58*, 137 (2007).
[52] Yariv, M. M. Rapport and L. Graf, *Biochem. J. 85*, 383 (1962).
[53] R. Butowt, A. Niklas, M. I. Rodriguez-Garcia and A. Majewska-Sawka, *J. Plant Res. 112*, 107 (1999).
[54] Y. Qin and J. Zhao, *J. Exp. Bot. 57*, 2061 (2006).
[55] L. Kimberly and E. A. Nothnagel, *Protoplasma 196*, 87 (1997).
[56] Wisniewska and A. Majewska-Sawka, *Plant Cell Rep. 26*, 1457 (2007).
[57] J. Van Hengel, A. B. Van Kammen and S. C. De Vries, *Physiol. Plant. 114*, 637 (2002).

[58] V. Dyachock, M. Wiweger, L. Kenne and S. Von Arnold, *Plant Physiol. 128*, 523 (2002).
[59] J. Sherrier, G. S. Taylor, K. A. T. Silverstein, M. B. Gonzales and K. A. VandenBosch, *Protoplasma 225*, 43 (2005).
[60] J. Kieliszewski and D. T. A. Lamport, *Plant J. 5*, 157 (1994).
[61] A. Rathbun, M. J. Naldrett and N. J. Brewin, *Mol. Plant Microbe Interact. 15*, 350 (2002).
[62] Schröder, J. Lisso, P. Lange and C. Müssig, *BMC Plant Biol. 9*:20. doi:10.1186/1471-2229-9-20 (2009).
[63] P. W, Becraft, *Trends Plant Sci. 3*, 384 (1998).
[64] Y. Matsubayashi, M. Ogawa, A. Morita and Y. Sakagami, Science *296*, 1470 (2002).
[65] Y. Matsubayashi and Y. Sakagami, *Proc. Natl. Acad. Sci. USA 93*, 7623 (1996).
[66] Ed. D. L. Schmidt, F. Guzzo, M. A. J. Toonen and S. C. De Vries, *Development, 124*, 2049 (1997).
[67] Albrecht, E. Russinova, B. Kemmerling, M. Kwaaitaal and S. C. De Vries, Plant Physiol. *148*, 611 (2008).
[68] P. Schellembaum, A. Jaques, P. Maillot, C. Bersch, F. Mazet, S. Farine and B. Walter, *Plant Cell Rep. 27*, 1799 (2008).
[69] De Veylder, T. Beeckman, G. T. Beemster, J. De Almeida Engler, S. Ormenese, S. Maes, M. Naudts, E. Van der Schueren, A. Jacqmard, G. Engler and D. Inze, *EMBO J. 21*, 1360 (2002).
[70] L. Dupuy, J. Mackenzie and J. Haseloff, *Proc. Nat. Acad. Sci. 107*, 2711 (2010).
[71] E. D. Shpak, C. T. Berthiaume, E. J. Hil and K. U. Torii, *Development 131*, 1491 (2004).
[72] Z. Magyar, L. De Veylder, A. Atanassova, L. Bako, D. Inze and L. Bogre, *Plant Cell 17*, 2527 (2005).
[73] J. Doonan, *Curr. Opin. Plant Biol. 3*, 482 (2000).
[74] P. S. Springer, D. R. Holding, A. Groover, C. Yordan and R. A. Martienssen, *Development 127*, 1815 (2000).
[75] Desvoyes, M. P. Sanchez, E. Ramirez-Parra and C. Gutierrez, *Heredity 105*, 80 (2010).
[76] Skirycz, H. Claeys, S. De Bodt, A. Oikawa, S. Shinoda, M. Andriankaja, K. Maleux, N. B. Eloy, F. Coppens, S. D. Yoo, K. Saito and D. Inzé, *Plant Cell* doi:10.1105/tpc.111.084160 (2011).
[77] Francis and D. A. Sorrell, *Plant Growth Regulation 33*, 1 (2001).
[78] R. P. Bhalerao and M.J. Bennett, *Nature Cell Biol. 5*, 939 (2003).
[79] Y. S. Kim and C. M. Park, *Plant Signal Behav. 2*, 15 (2007).
[80] Riou-Khamlichi, R. Huntley, A. Jacqmard and J. A. H. Murray, *Science 283*, 154 (1999).
[81] K. Zhang, L. Diederich and P. C. L. John, *Plant Physiol. 137*, 308 (2005).
[82] Y. Hu, F. Bao and J. Li, *Plant J.* 24, 693 (2000).
[83] Y. Matsubayashi, L. Takagi and Y. Sakagami, *Proc. Natl. Acad. Sci. USA 94*, 13357 (1997).
[84] Y. Amano, H. Tsubouchi, H. Shinohara, M. Ogawa and Y. Matsubayashi, *Proc. Natl. Acad. Sci. USA 104*, 18333 (2007).
[85] Kutschmar, G. Rzewuski, N. Stührwohldt, G. T. S. Beemster and D. Inzé, *New Phytol. 181*, 820 (2009).

[86] Y. Matsubayashi, M. Ogawa, H. Kihara, M. Niwa and Y. Sakagami, *Plant Physiol. 142*, 45 (2006).
[87] L. Song, Q. M. Shi, X. H. Yang, Z. H. Xu and H. W. Xue, *Cell Res. 19*: 864 (2009).
[88] E. Benson, P. T. Lynch and G. N. Stacey GN, *Agbiotech News and Information 10*, 133 (1998).
[89] Panis, H. Schoofs, S. Remy, L. Sagi and R. Swennen, Cryopreservation of tropical plant germplasm, Engelmann and Takagi (ed) JIRCAS, Japan, 212 (2000).
[90] Q. Wang, P. Li, U. Hanania, N. Sahar, M. Mawassi, R. Gafny, I. Sela, E. Tanne and A. Perl, *Plant Sci. 168*, 565 (2005).
[91] Mikula, K. Tomiczak and J. J. Rybczynski, *Plant Cell Rep. 30*, 565 (2011).
[92] Yokomuro, N. Shiono, T. Ozawa, T. Fujii, Y. Watanabe, N. Koyama and M. Okada, *Ann. Thorac. Cardiovasc. Surg. 16*, 105 (2010).
[93] W. McGinnis, M. S. Levine, E. Hafen, A. Kuroiwa and W. Gehring, *Nature 308*, 428 (1984).
[94] Volbrecht, B. Veit, N. Sinha and S. Hake, *Nature 350*, 241 (1991).
[95] Y. Zhao, Y. Hu, M. Dai, L. Huang and D. X. Zhou, *Plant Cell 21*, 736 (2009).
[96] Y. Hirakawa, Y. Kondo and H. Fukuda, *Plant Cell 22*, 2618 (2010).
[97] P. Nurse, *Nature 344*, 503 (1990).
[98] L. De Veylder, T. Beeckman, G. T. Beemster, L. Krols, F. Terras, I. Landrieu, E. Van der Schueren, S. Maes, M. Naudts and D. Inze, *Plant Cell 13*, 1653 (2001).
[99] A. Sorrell, M. Menges, J. M. Healy, Y. Deveaux, C. Amano, Y. Su, H. Nakagami, A. Shinmyo, J. H. Doonan, M. Sekine and J. A. Murray *Plant Physiol. 126*, 1214 (2001).
[100] R. E. Rodriguez, M. A. Mecchia, J. M. Debernardi, C. Schommer, D. Weigel and J. F. Palatnik, *Development 137*, 103 (2010).
[101] Wang, T. Soyano, S. Machida, J. Y. Yang, C. Jung, N. H. Chua and Y. A. Yuan *Genes & Dev* 14. doi: 10.1101/gad.1985511(2010).

Reviewed by:

Prof. Thierry Wetzel (RLP-AgroScience.GmbH, Germany)
thierry.wetzel@agroscience.rlp.de.
Dr. Samia Daldoul (Center of Biotechnology, Tunisia)
samiabiotech@gmail.fr.
Prof. J. Paul Knox (Center of Plant Sciences, Leeds, UK)
j.p.knox@leeds.ac.uk.

In: Cell Proliferation
Editors: C. Zhang and X. Zeng
ISBN: 978-1-62417-352-3

Chapter VI

Cell Proliferation in Drug Discovery and Development

Gopalan Soman**, *Xiaoyi Yang, Steve Giardina and George Mitra
Biopharmaceutical Development Program, SAIC Frederick, Inc.,
National Cancer Institute, Maryland, US

Abstract

Cell proliferation is one of the major processes controlling the states of life and disease in all living organisms. Measurements of cell proliferation, proliferation inhibition, cytolysis, and apoptosis are important for drug discovery, development and clinical evaluation. Cell proliferation is regulated by a variety of cellular factors including growth factors, cytokines, and cytotoxins. Drug treatment affects the cell's proliferative response through different signal transduction pathways. *In vitro* cell proliferation assays, together with *in vivo* growth assays, are widely used in laboratories for basic research and pre-clinical drug evaluation studies. High-throughput screenings, and cellular and drug specificities are critical elements in successful drug discovery efforts. As drug development efforts progress, more attention is paid to assay reproducibility, consistency and variability. The molecular mechanism-oriented cell proliferation assay is critical for demonstrating the target specificity of new drugs under development.

In this chapter, we focus on the uses of non-radioactive cell proliferation assays, including the relative metabolic rate, or relative activity measurements using MTS dyes for cell proliferation assays for cytokines, such as interlukins-7 (IL-7) and cytokine derivatives; cytotoxicity measurement of immunotoxins, such as MR1-1, which is a disulfide-stabilized variable domain immunotoxin (the catalytic domain containing Pseudomonas exotoxin) that targets EGFRviii, a mutant form of the epidermal growth factor receptor (EGFR) and cytokine-Fc conjugates; and differential cell-killing assays of therapeutic recombinant viruses, for instance, adenoviral vectors (such as Ad5-D24-RGD and Ad5/3-D24). We also demonstrate the application of a recently introduced Real Time Cell Electronic Sensing (RT-CES) method in a label-free cell proliferation assay and its correlation with MTS-based cell proliferation assays for assessing the activity of an

* E-mail address: somang@mail.nih.gov.

immunotoxin (MR1-1) and differential cytolysis of the oncolytic adenoviral vector Ad5-D24-RGD. Recently, reporter gene expression assays have been effectively applied to monitor cell proliferation correlating to reporter gene expression. A signal transducer and activator of the transcription 3 (STAT3) luciferase reporter gene expressed in HeLa cells was used to show the correlation of the STAT3 gene expression to cell proliferation monitored by an MTS assay. Cell-based bioassays generally display significant variations in the absolute activity defined by the concentration of drug products that displays half-maximal stimulatory or half-maximal inhibitory responses (ED_{50}). The relative activity of a test article is defined by the ratio of the activity, or the ED_{50} value, compared to that of the reference standard, showing reduced intra- and inter- day variations. The assay variations under optimized and qualified assay conditions are also discussed. The stringent control of cell growth, and the maintenance and quality of the cells frequently becomes critical for the successful use of cell proliferation assays for product release and stability monitoring, as demonstrated here with the IL-7-induced cell proliferation of pre-B cells in the culture.

1. Introduction

Cell proliferation is regulated by a variety of cellular factors such as growth factors, cytokines, cytotoxins, etc. Drug treatment produces effects on the cell's proliferative response through different signal transduction pathways. Cell proliferation plays a critical role in the normal development, growth, differentiation, tissue homeostasis, etc. Disruptions of cell division, proliferation or cell death represent the underlying causes of all forms of cancers, and contribute to the pathology of birth defects, degenerative diseases and many other anomalous conditions underlying several diseases. Understanding the mechanistic pathways that contribute to the disruption of the normal proliferation or growth of a cell is essential to better understanding health and disease in all forms of life. Several of the methodologies that can be applied to the cell cycle and growth analysis in Zebra embryos are summarized in a recent article [1]. Profiling key cellular indicators, such as apoptotic fraction and stage of apoptosis, viability, cell cycle, cell counts, and transfection efficiency, or target gene expression levels, cell death and cytotoxicity are commonly used to understand the cellular mechanisms involved in drug efficacy and toxicity.

Various methodologies for performing assays *in vitro* to measure cell viability, cell growth or proliferation are outlined in a number of recent reviews [2-5]. These include the classical methodologies such as dye exclusion and direct microscopic counting, measuring DNA content using propidium iodide labeling and flow cytometry, DNA synthesis following tritiated thymidine or 5-bromo-2'-deoxyuridine (BrdU) uptake, measuring the metabolic rate by using ATP quantitation reagent or the redox dye MTT and its derivatives or analogues, and so on. Apoptosis assays, using luminescent caspase detection, as well as reporter gene assays have emerged during the last decade. More recently, non-invasive biosensor technologies, such as electrical cell-substrate impedance sensing (ECIS) and real time cell electronic sensing (RT-CES), have been introduced to dynamically monitor cell proliferation [6, 7]. These *in vitro* cell proliferation assays, together with *in vivo* growth assays, are widely used in laboratories for basic research and pre-clinical drug evaluation studies. The transition from the classical radioactive [^{3}H] thymidine incorporation to alternative cell-based assays is necessitated by the need for avoiding radioactive material as well as the requirement for rapid high-throughput screening assays for basic research and drug discovery research.

Cell proliferation assays are currently core development tools, and various analytical laboratories provide assay development and quality control studies under contract. High-throughput screening (HTS), and understanding cellular and drug specificity are critical elements in any successful drug discovery effort. The choice of methodology at the development stage is determined by the speed and ease of the assay as well as by the HTS output rate and automation capability. An obvious approach to HTS for anticancer activity is to look for compounds that reduce the growth of cancer cells in cultures. Many cell lines that have been derived from a wide variety of human cancers are available. The effects of compounds on such cells can be compared to effects on non-cancerous cells to establish if there is any selectivity for tumor-derived cells [8]. The National Cancer Institute's (NCI) 60 human tumor cell line anticancer drug screen (NCI60) is one example of a rapid *in vitro* cell screen for the identification of cellular targets and is a mechanism of anti-cancer drugs in the drug discovery effort [9]. A study by Iljin et al. [10] represents a recent example of the successful use of cell-based screens. In this study, the researchers used a panel of prostate cancer cells and non-cancerous prostate epithelial cell lines. In another study, the marine-derived compound batzelline was shown to be selectively cytotoxic to pancreatic cell lines over the normal kidney epithelial cell line Vero [11]. The complexity inherent in the process of cell proliferation and cell death necessitates the careful choice of cell-based assays for measuring these processes. An assay must be validated to meet the requirement of the study. A non-exhaustive comparison of methods to detect cell death with apoptotic or non-apoptotic morphologies, including their respective advantages and pitfalls, is summarized in a recent review [2]. The cytotoxicity assays used in HTS need to quantify the extent of cell death, and must also be able to distinguish between the various pathways of cell death (mechanism based). The approaches used to quantify distinct cell death pathways, and their advantages and pitfalls are summarized in a recent review article [12]. The usage of proliferation assays very often demands that related assays be performed to verify findings or to provide additional information. Two critical factors that contribute to successful drug discovery and screening are the assays used to characterize the drug, proposed mechanism of action, and the drug specificity. Knowledge of the purity and identity of the drug molecules is essential especially for cell proliferation or related cell-based assays, as the buffer components, impurities in the product, and the conjugation reagents or excipients used for targeted delivery etc. may influence the assay and hence the outcome. Many of the anti-cancer peptides and oligonucleotides frequently result in misleading conclusions regarding the drug action and specificity, and such an incorrect conclusion can be costly. As the drug development efforts progress, more attention is paid to the assay's reproducibility, consistency and variability. Key potency or efficacy marker assays must be selected and qualified to characterize and define the potency index for the drug before entering clinical investigation. The choice of assay at this stage depends primarily on the assay parameters, the reproducibility of results within laboratory and inter-laboratory tests and the relevance of the assay's biological or functional activity in relation to the expected clinical potency or efficacy.

Cell-based bioassays generally display significant variations in the absolute activity defined by the concentration of the drug product that displays half-maximal stimulatory or half-maximal inhibitory responses (ED_{50}). The relative activity of the test article, defined by the ratio of the activity of the test article compared to that of the reference standard, shows much improved intra- and inter-day variability. The assay variability under optimized and qualified assay conditions should also be reviewed as it is very critical for potency assays for drug products in clinical development. However, the ED_{50} values alone cannot be used to predict whether different lots of the drug preparation are equivalent or comparable. The dose-response curves should be parallel in order to ensure the validity of relative activity measurement. The stringent control of cell growth, maintenance and checks on the quality of the cells are critical for successfully using cell proliferation assays with product release and stability monitoring, as demonstrated here with the IL-7-induced cell proliferation of pre-B cells in the culture.

2. The Application of MTS Colorimetric Assay for Measurement of Cell Proliferation and Cytotoxicity

2.1. MTS Colorimetric Assay Introduction

The CellTiter 96® AQ_{ueous} assay uses the novel tetrazolium compound (3-(4, 5-dimethylthiazol-2-yl)-5-(3-carboxymethoxyphenyl)-2-(4-sulfophenyl)-2H-tetrazolium, inner salt; MTS) and the electron coupling reagent, phenazine ethosulfate (PES). MTS is chemically reduced by metabolically active cells into formazan, soluble in a tissue culture medium. The measurement of the formazan's absorbance can be carried out using a microplate reader at 490nm [13]. The assay measures dehydrogenase enzyme activity found in metabolically active cells. Because the production of formazan is proportional to the number of living cells, the intensity of the color produced is a good indicator of the viability of the cells [14]. Elements of the qualification of this assay and the standardization of reference material for cell proliferation assays were described earlier for a recombinant human IL-15 (rhIL-15)-induced CTLL-2 (a clone of cytotoxic T lymphocyte cell line, ATCC Cat#TIB-214) cell proliferation [15].

The MTS colorimetric cell proliferation assay is performed following a procedure similar to that described earlier for rhIL-15 [15]. Cells are seeded in a 96 well plate, incubated for 24 hours at 5% CO_2 at 37°C. The stimulators or cytotoxin/cytolytic agents (viral vectors) are added and incubated for a pre-determined time period. A CellTiter 96® AQ_{ueous} reagent is added and incubated for an additional 3 to 4 hours. The plates are read at 490 nm. Data is analyzed by a four-parameter curve-fit analysis using SoftMax Pro software from Molecular Device (currently GE Healthcare). The four-parameter curvefit analysis uses the equation $Y \sim \{(A - D) / [1 + (X / C)^B]\} + D$ where Y = the response, optical density (OD); X = the arithmetic dose, concentration in ng/ml; A = the response when X = 0, the lower asymptote; D = the response when X is "infinite", the upper asymptote; C = the X value, concentration, resulting in a response halfway between A and D, referred to as ED_{50}; and B = a "slope" factor [15]. Assays are optimized to meet the requirements of intended use, and a standard procedure is followed for routine regulated (GxP) tests.

2.2. Cytokines and Cell Proliferation

Cytokines are a family of molecules that include the inflammatory cytokines, such as the interleukins and interferons; growth factors, such as epidermal and hepatocyte growth factors; and chemokines such as the macrophage inflammatory proteins [16]. Cytokines, in general, play a critical role in the development of lymphoid tissues as well as the ultimate differentiation of naive and memory T-cells [17].

Voluminous reports may be found in the literature on the cytokine stimulation and cellular specificity of cell growth. While most of the *in vitro* cell proliferative data is generated using the [^{3}H] thymidine uptake (DNA synthesis), or other radiolabeled nucleus, metabolic activity measurements using MTT or MTT analogues have emerged recently as an alternative, non-radioactive colorimetric assay [18]. Because of the interdependence and pleotropism of various cytokines, the plasma and other body fluid concentration measurements of cytokines are problematic. Clinical interpretation of the concentration of any single cytokine measured is not reliable. For any meaningful deduction of an agent's toxicity, multiple cytokines should be measured; cell proliferation assays for individual cytokines will be meaningless because different cytokines may induce a proliferative response on the same cell. However, the assay specificity could be achieved to certain degree by using neutralizing antibodies specific for interfering cytokines or growth factors. A variety of ELISA-based multiplex cytokine assays are currently available for determining the levels of multiple cytokines in plasma and other body fluids [19-22]. Cytokine-based therapy is a very attractive approach. Several cytokines or derivatives have been successful in the clinic as biotherapeutics [23].

The measurements of the biological activity of the cytokine and its associated stability over time are critical factors in the clinical development effort. *In vitro* cell proliferation assays using susceptible cells are frequently used as surrogate potency assays. The available methods for *in vitro* potency assays and associated validation requirements were reviewed by Mire-Sluis, et.al. [24] and Meager [25]. A qualified MTS dye-based colorimetric assay method using the CellTiter 96® AQ_{ueous} colorimetric assay for the IL-15 stimulation of CTLL-2 cells was reported recently [15].

2.2.1. MTS Colorimetric Cell Proliferation Assay for Interlukin-7

Interlukin-7 (IL-7) is a member of the double-helical bundle-structured cytokines that include IL-2, IL-15 etc., sharing a common cytokine receptor along with IL-2, IL-4, IL-15 and IL-21 [26, 27]. IL-7 was originally characterized as a hematopoietic growth factor that was able to stimulate the proliferation of lymphoid progenitors [28]. Subsequently, IL-7 was shown to have a role in the T-cell growth and survival of both memory and naive-T-cell populations [29-31]. Recently, more functions have been reported, for instance, the involvement of IL-7 in the development of the intrinsic function of marginal zone B cells and the marginal microenvironment [32], the up regulation of cyclin D1 to promote lung cell proliferation [33], the recruitment of monocytes/macrophages to the endothelium [34], the mediation of Ebf-1 (early B cell factor-1) dependent lineage restriction in early lymphoid progenitors [35] and contributions to the progression of human T-cell acute lymphoblastic leukemias [36]. Opposing functions of IL-2 and IL-7 in the regulation of immune responses are also highlighted in a recent article [37].

Preclinical and clinical evaluation studies on IL-7 are summarized in a recent review [38]. The administration of recombinant human IL-7 (rhIL-7) to normal or lymphopenic mice, non-human primates and humans resulted in widespread T-cell proliferation and increased T-cell numbers, modulation of peripheral T-cell subsets, and increased T-cell receptor repertoire diversity. Preclinical studies suggest that IL-7 has both an immune restorative effect and a vaccine adjuvant effect, and may have beneficial effects in the setting of adoptive cell therapy. A number of clinical investigation studies have been conducted using rhIL-7 produced in *E. coli* with a limited number of patients, and the studies suggest that IL-7 therapy is safe, well tolerated and produces a potent immune restorative effect.

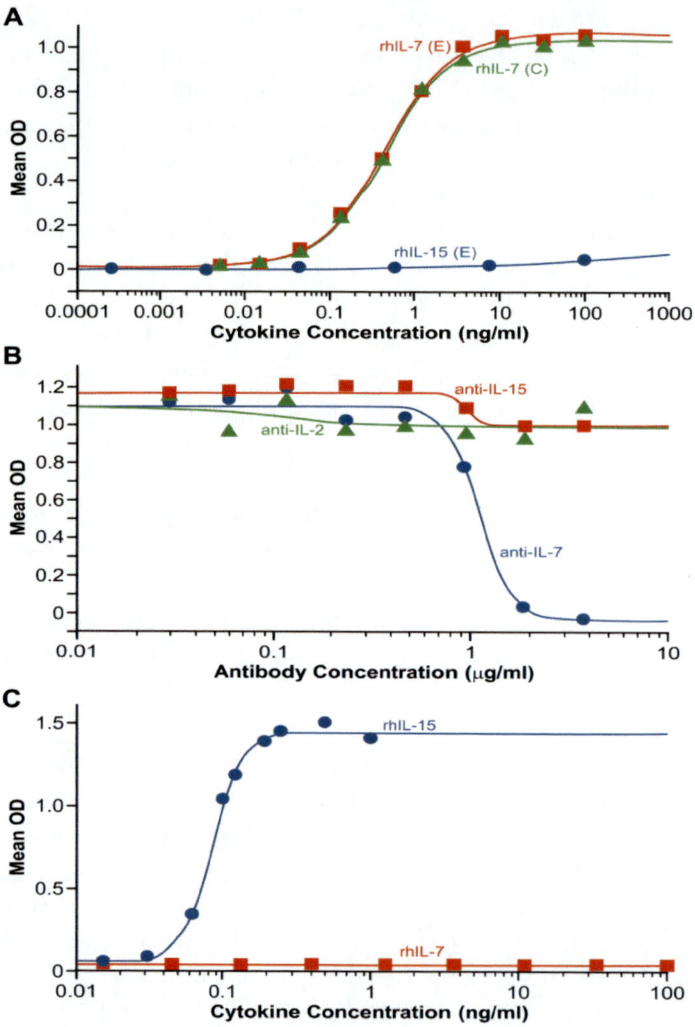

Figure 1. IL-7 stimulates specifically the murine pre-B cell 2E8 proliferation.
A: rhIL-7 produced in E.coli (E) and Chinese Hamster Ovary (CHO) cell (C) expression systems stimulates 2E8 cell proliferation but rhIL-15 expressed in E.coli does not show this effect; B: Anti-IL-7 antibody neutralized (blocked) IL-7 induced 2E8 cell proliferation, while anti-IL-15 and anti-IL-2 antibodies have no effect; C: rhIL-7 does not stimulate CTLL-2 cell proliferation under conditions of rhIL-15 stimulation.

IL-7 is a molecule of high interest in clinical evaluation for immunotherapy and as a vaccine adjuvant. Multiplex cytokine module assays are useful in determining levels of IL-7 in serum and body fluids. IL-7 plays a crucial role in the proliferation of pre-B lymphocyte cells. A number of pre-B cell lines have been recently established and shown to proliferate in the presence of IL-7 [39-46].

The stimulatory effect of pre-B cells with IL-7 is shown to be very specific [40, 44]. The murine pre-B cell has been shown to be stimulated with both murine and human IL-7 [39, 42, 44, 45, 47, 48], using a [^{3}H] thymidine incorporation assay. A cell proliferation assay using the murine pre-B cell line (PB1) has been optimized and qualified with respect to cellular and cytokine specificity and consistency [44]. The cytokine specificity of 2E8 (a murine pre-B cell, ATCC Cat #TIB-293) cell proliferation was also demonstrated using [^{3}H] thymidine incorporation as well as by MTT assays [40]. A typical IL-7 dose-response curve of 2E8 cell proliferation as measured by the MTS colorimetric assay is shown in Figure 1A. In contrast, a related cytokine IL-15 does not stimulate the 2E8 cells in the identical dose-range. This is consistent with the reported specificity of the IL-7 stimulation of pre-B cells [40, 44].

Table 1. Maximal dose-response read out and Cell viability dependence on Cell Passage

Cell passage #	Maximal OD 490 nm	Cell viability
1	0.624	66
2	0.549	71
3	1.893	83
4	1.353	76
5	1.509	78
6	1.497	77
7	1.529	76
8	1.245	78
9	1.300	75
10	1.300	74
11	1.320	70
12	1.290	68
13	0.897	69
14	1.251	68
15	1.000	69
16	1.524	60
17	0.912	50
18	1.230	46
19	0.654	42

IL-7 dependent 2E8 cell proliferation is inhibited (blocked) by IL-7 antibodies, but not by antibodies of IL-2 or IL-15, which establishes the specificity of IL-7 (Figure 1B). IL-7 responses to T-cells and other cell lines are well established [49]. The rhIL-7 did not show any stimulation of the CTLL-2 cell line at the concentration range where rhIL-15 shows its maximal response (Figure 1C). The dose-response curves and curve fit parameters for susceptible cells depend on the cell growth characteristics as well as the cell passage number and cell viability [15, 24, 25, 50]. The 2E8 cell growth in cultures is largely dependent on the cell passage cycle (Table 1). Cell growth was monitored by measuring the maximal absorbance reading from the MTS colorimetric assay as determined by the upper asymptote of the dose-response curves. Consistent with the reduction in cell growth, there is a parallel decrease in cell viability as determined by the microscopic cell count using Typan blue. Cells in the range of passage 3-11 show a higher viability (>70%). The cell viability drops significantly after passage 11. Despite apparently comparable ED_{50} values with cells at different passages (Figure 2), the dose-response curves are not parallel and hence the relative response of the cells at different passages cannot be assessed based on ED_{50} values only.

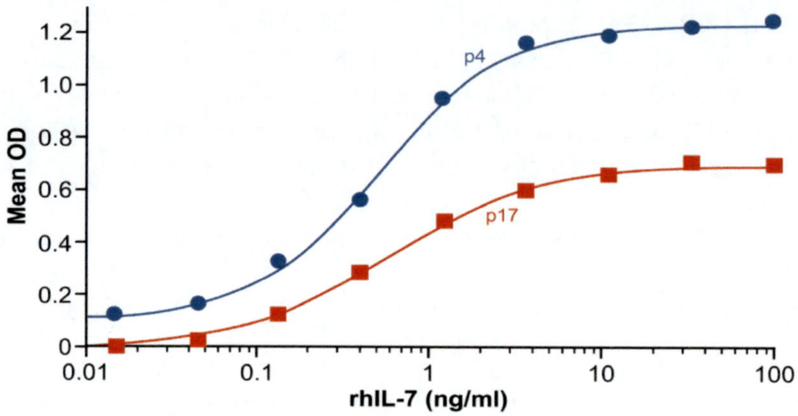

Figure 2. The rhIL-7 dependent dose-response curves for 2E8 cells are dependent on the cell passage number or cell growth characteristics (p.4 and p17 are passage 4 and passage 17 in table 1 respectively).

2.2.2. Assay Consistency and Variability

An important aspect of bioassay development for the clinical product release and stability testing is the consistency and/or variability of the assay within one plate and between plates, both intra-day and inter-day. Inherent assay variability in bioassays is generally the rule rather than the exception.

Table 2. Intra-plate and within day inter-plate variations

Plate #	ED_{50}		Relative activity*
	Region 1	Region 2	
1	0.632	0.626	1.010
2	0.535	0.616	0.869
3	0.624	0.635	0.983
Plate Average	0.597	0.626	0.954
Std.Dev.	0.054	0.010	0.075
Coefficient of variability (CV%)	9.019	1.519	7.855

* Relative activity = ED_{50} region 1 ÷ ED_{50} Region 2.

The relative activities could also be determined by using a defined reference standard to generate a standard curve and analyzing the test sample at different dilutions. Test sample read-outs (only those read out within the linear part of the standard curves are considered valid) are extrapolated to the standard curve to calculate the relative concentration [15]. Generally, day-to-day variations are observed in all the curve-fit parameters.

Table 3. Intra-day inter-plate and inter-day variations in the ED_{50} values

Date	Cell Passage	Viability %	Sample	Activity	Plate #1	Plate #2	Plate #3	Mean	Std. Dev.	CV%
6/9/ 2011	p3	80%	ref lot	ED_{50}	0.822	0.797	0.824	0.814	0.015	1.84
			test lot	ED_{50}	1.081	1.054	1.049	1.061	0.017	1.60
				RA*	76%	76%	76%	76%	-	-
6/15/ 2011	p5	78%	ref lot	ED_{50}	0.642	0.640	0.672	0.651	0.018	2.76
			test lot	ED_{50}	0.824	0.820	0.854	0.833	0.019	2.28
				RA*	78%	78%	79%	78%	0.58%	0.74
6/22/ 2011	p8	69%	ref lot	ED_{50}	0.994	1.098	0.905	0.999	0.097	9.71
			test lot	ED_{50}	1.267	1.451	1.270	1.329	0.105	7.90
				RA*	78%	76%	71%	75%	3.61%	4.81

* RA- relative activity = ED_{50} ref lot ÷ ED_{50} test lot.

The results of MTS-based, pre-B cell 2E8 proliferation by IL-7 shows the similar type of assay variability that is observed with other cytokines such as IL-2, IL-15 or cytokine conjugates. However, the maintenance of 2E8 cells in the continuous culture was more difficult. Although CTLL-2 cells could be maintained in an IL-2 supplemented culture medium for more than 50 cycles, 2E8 cell viability dropped significantly in the murine IL-7 supplemented culture medium after 10-11 passages. The cell growth may be better controlled by the optimization of the IL-7 concentration in the culture. The pre-B cell line PB1 was shown earlier to be at least 10 fold more sensitive to IL-7 stimulation than 2E8 [44].

Because of the inherent assay variations within the lab and between labs, bioassay results are generally expressed as relative activity to a well-defined reference standard material. Whenever international standards are available, the in-house reference standard needs to be standardized using the international standard and the activity is to be expressed in international units based on the international standard [15, 25]. The approach used in our laboratory is described in an earlier publication on IL-15 [15]. The standardization using an international reference material is performed after the assay is optimized and assay variations are established.

Intra-plate and Inter-plate Variability

Because the cell proliferation assay is a relative, rather than an absolute method, the most important aspect of assay reliability is related to the "within-plate variability" (regional effects), as both the product and reference material are run within the same plate. Dose-response curves of cell proliferation assays generally show significant intra- and inter-assay variations in their curve-fit parameters.

The absolute activity generally defined by the ED_{50} values (concentration of the ligand or product at the half-maximal stimulation) shows a much higher variation. The dose-response curves in a typical cell proliferation or proliferation inhibition assay are usually sigmoidal, and a four-parameter curve-fit or Log-Logit curve fit is performed in many laboratories, although alternate modes of curve-fit and data analysis are also frequently used. The C parameter values of the four-parameter curve is considered as the half maximal dose-response concentration, and the ratio of the average C values of test sample to the C value of the standard is reported as the relative activity. While there is notable intra-plate regional variation in the ED_{50} values, the ED_{50} ratio between regions is nearly 1, with a minimal variation (Table 2).

The assay was performed in three different plates. The same material was used in serial dilutions in hex plicate wells. The six well regions were split into two sets of triplicate well regions (Region 1 and Region 2). Region 1 is considered as a reference and region 2 as a test material. The theoretical relative activity is 1.0 as both test and reference are identical. Considering the two regions in triplicate plates are identical, the hex plicate values for ED_{50} gives an average of 0.611 ± 0.038 ng/ml with a CV 6.2%.

Intra-day and Inter-day Variability

Wider variations in the curve-fit parameters were noticed during the initial stages of assay development, and there is better consistency after assay optimization and operator training.

Inter-day variability in the assay parameters occurs with cells cultured at different times independent of passage number, whereas variability is much less for the within-day inter-plate (Table 2 and 3). However, the relative activity expressed as the ratio of ED_{50} values of a standard and test article shows much less variation (Table 3). The inter-day variations in the individual ED_{50} values were 22.2% for Lot 1 and 22.7% for Lot 2, whereas the variation in relative ED_{50} values was only 2%.

2.3. Immunotoxins and Cell Proliferation

Recombinant immunotoxins are antibody-toxin chimeric molecules that kill cancer cells via binding to a surface antigen, internalizing and delivering the toxin moiety to the cell cytosol [51-53]. In the cytosol, toxins catalytically inhibit a critical cell function and cause cell death. Immunotoxins are considered to have a potential as future drugs for cancer immunotherapy. There are a large number of comprehensive reviews on the use of cytotoxins and immunotoxins in cancer therapy [54-68].

So far, only one immunotoxin (Denileukin diftitox) is approved by the US Food and Drug Administration. Immunotoxins targeted to the B-lymphoma specific-cell surface antigen CD22 (cluster of differentiation-22) are extensively evaluated in the clinic against hairy cell leukemia [62, 63, 69]. These include BL22 (a CD22-directed ADC, representing a group of single-chain immunotoxins with scFv fragments fused to a truncated version of *Pseudomonas aeruginosa* exotoxin A) and HA22 (a modified version of BL22, with a higher affinity among others) [69].

MR1-1, BL22 and HA22 are a few of the immunotoxins generated and evaluated in the laboratory of the Molecular Biology of National Cancer Institute led by Dr. Ira Pastan [69-71]. BL22 showed high response rates in phase II clinical trials in patients with hairy cell leukemia, achieving up to a 47% complete remission rate [61, 62].

Determining the specific cytotoxicity of an immunotoxin on target cells is a key to evaluating its therapeutic potency. Before recombinant immunotoxins can be evaluated for anticancer activity in humans, they undergo extensive preclinical testing. Immunotoxins must demonstrate cell-killing activity in the tissue culture, antitumor activity in an animal model and have favorable pharmacokinetic and toxicity profiles.

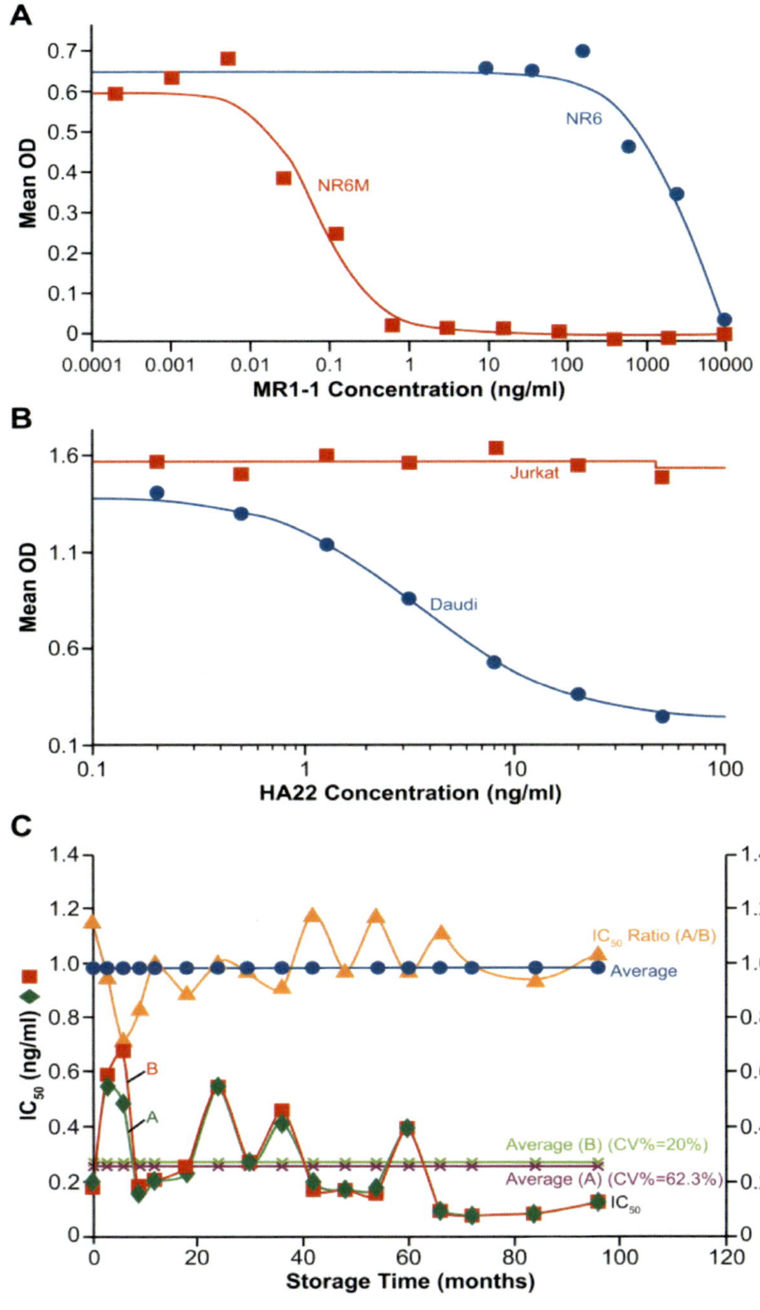

Figure 3. Targeted cell specific cytotoxicity of MR1-1 and HA22 and stability of MR1-1. A: MR-1 shows cytotoxicity to a EGFRvIII gene-transfected NR6M cell with no effect on the control-non-transfected parent cell NR6; B: HA22 shows targeted cell specific proliferation inhibition on the CD22 expressing B lymphoma Daudi cells but no effect on the CD22 lacking Jurkat cell; C: Stability of MR1-1. ED_{50} values of MR1-1 specific inhibition of NR6M cells shows wider inter-day variation compared to the relative activity expressed as the ED_{50} ratio of the Standard lot to test article.

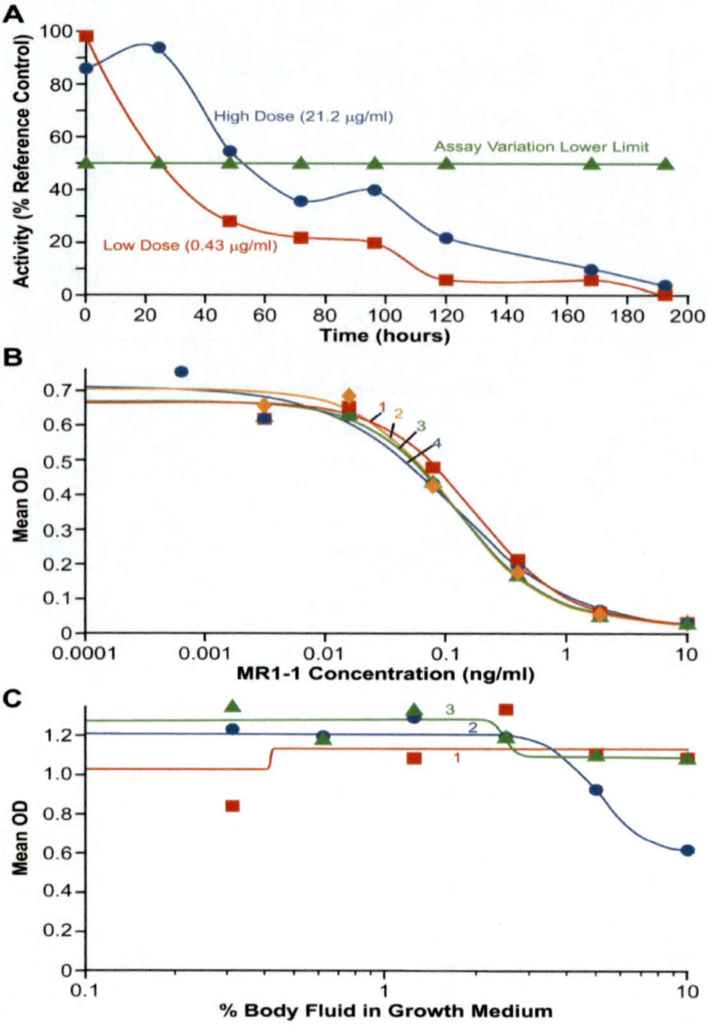

Figure 4. MR1-1 cytotoxicity tests in body fluids and clinical administration infusion model. A: Stability of MR1-1 under Clinical administration mock infusion conditions; B: 2% rat serum and CSF have no effect on the MR1-1 cytotoxicity; MR1-1 was diluted in 2% serum (1) or CSF (3) to an initial concentration of 10ng/ml. Subsequent dilutions from the 10ng/ml are performed in a control medium. Control (standard) is the same MR1-1 Lot diluted in the medium throughout (4) or the medium containing 2% rat serum (2). There is no difference in the titration curves suggesting that 50- fold dilutions of samples can be assayed without significant interference; C. Direct effects of body fluid on NR6M cell growth. 1: Control regular culture medium without body fluid; 2: Culture medium containing rat serum (0.5 to 10%); 3: Culture medium containing rat CSF (0.5 to 10%).

2.3.1. MTS Colorimetric Cell Proliferation Assay for Immunotoxins

Most of the immunotoxins are cytotoxic to their target cells. Cell proliferation inhibition or cell-killing assays using targeted cells are used for defining the functionality and specificity of immunotoxins. The [^{3}H] leucine protein synthesis assay has been used in several laboratories to determine the cytoxicity of different immunotoxins [69, 70, 72]. MR1-1 is an immunnotoxin targeted to cells bearing an epidermal growth factor receptor (EGFR) on cell surfaces [70, 73-75]. This immunotoxin is extensively investigated and currently in clinical evaluation. BL22 is an immunotoxin targeted to CD22-expressing lymphoma cells and is one of the most extensively studied immunotoxins for the treatment of hairy cell leukemia [71].

HA22 is an improved form of BL22 in which the Fv was mutated via an antibody phage display to isolate a mutant phage that improved binding to CD22. In HA22, residues of Ser-Ser- Tyr in the heavy-chain CDR3 (complementarity determining region 3) have been mutated to Thr-His-Trp [69].

HA22 has a 5 to 10-fold increase in cytotoxic activity on various CD22-positive cell lines and is up to 50 times more cytotoxic to cells from patients with chronic lymphocytic leukemia and hairy cell leukemia. The cytotoxicity and cytolytic activity of these drug molecules was also studied using a $l^3H]$ leucine incorporation assay and colorimetric assays using WST-1 [2-(4-Iodophenyl)-3-(4-nitrophenyl)-5-(2,4-disulfophenyl)-2H-tetrazolium] and WST-8 [2-(2-methoxy-4-nitrophenyl)-3-(4-nitrophenyl)-5-(2,4-disulphonyl)-2H-tetrazolium] as well as MTS [69, 70, 72]. MR1-1 was shown to inhibit the proliferation of a NR6M (EGFR-transfected Glioma cell NR6) and EGFRvIII-transfected U87MG astrocytoma cells (ATCC Cat#HTB-14) with no effect on the non-transfected parent cells [70].

The EGFRvIII-transfected NR6M cell proliferation was effectively inhibited by MR1-1 with an IC_{50} ~ 0.1 ng/ml range, whereas the normal non-transfected control cell, NR6, showed little or no inhibition with MR1-1 up to 50 ng/ml (Figure 3A). Under identical conditions, neither the VH (variable region of heavy chain) of the MR1-1 nor another recombinant single-chain Fv-ERB2 (variable chain antibody fragment domain-estrogen receptor-2) conjugate showed any inhibition of NR6M cell proliferation. BL22 and HA22 showed to inhibit the proliferation of CD22-expressing B lymphoma cell lines Raji, Daudi and CA46 (a Burkitt lymphoma cell line), but had little or no effect on the CD22 negative Hut102 (a human cutaneous T-lymphocyte cell line, ATCC Cat #TIB-162) and L540 cells (a Hodgkins-derived cell line) [69].

The MTS colorimetric assay also showed that HA22 is cytotoxic to Daudi cells with little or no effect on the T-cell line Jurkat (Figure 3B). Two lots of MR1-1 manufactured under current Good Manufacturing Practice (cGMP) conditions, formulated in phosphate-buffered saline and stored at -80°C freezer were tested for activity for 8 years (Figure 3C), following an optimized standard procedure using an MTS colorimetric method. Similar to other cell proliferation and cell proliferation inhibition assays, the MR1-1-induced cell proliferation inhibition also showed significant inter-day variability in absolute activity as defined by the IC_{50} values (Figure 3C).

The CV of the average ED_{50} or IC_{50} for the 17 time point estimates over a 9 year period was 62.4% for the reference standard (Lot A) and 70% for the clinical test lot (Lot B). However, the relative activity expressed as the ratio of the IC_{50} of the reference standard and test article showed much smaller variations compared to the absolute IC_{50} values.

The relative activities of the test article ranged from 0.72 (72%) to 1.18 (118%), compared to the reference standard lot with an average of 0.984 (98.4%) and a standard deviation of 0.12 (CV 12.2%). The results clearly showed high inter-day variation in the absolute activity (IC_{50} values), but the relative activity of the test article remained comparable. A similar conclusion was deduced from experimental results on Daudi cell proliferation inhibition by HA22.

2.3.2. Application in Product Infusion Mock Studies

Apart from determining the potency of product lots and storage stability, the stability of a product under conditions of clinical administration is a very critical aspect of clinical development. The initial clinical administration studies were designed to have the product in the solution using an osmotic pump device for continuous infusion. An initial mock study using MR1-1 in saline at a low concentration revealed a complete loss of activity during the infusion period. Human albumin (0.2%) was used as an additive in the infusion formulation, and the mock infusion study was repeated at two concentrations of MR1-1 (Figure 4A). At a higher dose (21.2μg/ml), MR1-1 retained activity for 48 hours (>50% of control) within the limits of assay variation; and at a lower dose (0.43μg/ml), MR1-1 retained the activity within the limits of assay variation for 24 hours. A gradual loss of activity (time dependent) was observed, and the loss was significant after 24 hours of incubation for the low dose and after 48 hours of incubation for the high dose.

The results of these experiments highlight the importance of a controlled mock study of product activity and stability during the clinical administration, mimicking the actual clinical administration process as closely as possible. Such controlled studies are essential to ensure the functionality of drug products during the clinical administration. Similar mock studies are currently performed for all NCI sponsored pre-clinical and clinical studies.

2.3.3. Application in Pharmacokinetics

Another important aspect of drug development in pre-clinical and clinical evaluations is associated with the pharmacodynamics, pharmacokinetics and the concentration of drug molecules in tissue and body fluids.

The quantification of the functionally active drug molecule in tissues and body fluids is a challenging issue because the cellular components may interfere with the assay, either as agonists or antagonists of the drug action. Controlled studies to assess the level of interference and reliable sample dilution range for accurately determining the active drug molecules have to be worked out before the pre-clinical or clinical evaluation studies are performed. MR1-1 was shown to be stable in a human serum at 20g/ml [70]. With MR1-1, a pre-clinical safety study in support of the clinical study was performed in a rat model. The cytotoxicity assay was evaluated to assess the effect of body fluids (i.e., serum and cerebral spinal fluid [CSF], in this case).

Rat serum >5%, added to the routine assay medium, apparently has an inhibitory effect on NR6M cell proliferation (Figure 4B), suggesting that serum sample dilution < 20 X in the assay medium is inappropriate for accurately estimating MR1-1 when using this assay. A 50 X dilution of rat serum or CSF (Figure 4C) in the assay medium had no effect on the cytotoxicity (2% body fluid in the assay).

Further studies showed that the dose-response curves and IC_{50} values of MR1-1 in NR6M cell proliferation was not affected by the pre-dilution of the sample in 4% rat serum or 4% rat CSF (final concentration: 2% body fluid in the assay), suggesting that accurate estimates of the active MR1-1 concentration can be obtained when the samples are diluted ≥ 50 X in the assay medium. These studies highlight the importance of assay qualification studies prior to clinical and pre-clinical testing for reliable results.

2.4. Targeted Cell-Selective Cytolysis and Gene Therapy

During last two decades there has been rapid expansion in the field of gene therapy [76]. Viral vectors are the most effective means in current use for delivering target genes to humans. Nearly 30% of the clinical protocols reported use of adeno viral vectors or adeno-associated vectors. There was an evolution from the use of replicating defective to conditionally replicating adenoviruses (CRAd), which allowed preferential replication in cancer cells. *In vitro* cytolytic activity measurements for demonstrating targeted cell specificity are receiving more interest as a primary evaluation of targeted replication and cytolysis prior to safety evaluations in humans [77]. As an alternative, pre-clinical approach to animal testing, human non-prostate primary cells in *in vitro* cell cultures were used to test whether Ad[I/PPT-E1A] (a complex chimeric promoter which combines the T-cell receptor γ-chain alternate reading frame protein [TARP] promoter, with the prostate-specific antigen [PSA]- and PSMA- [prostate-specific membrane antigen] enhancer elements) replication is limited only to the prostate target cells by assessing the specificity and sensitivity of Ad[I/PPT-E1A] replication and cytolytic activity *in vitro* [77]. The MTS-based cell proliferation assay is used in several laboratories to demonstrate the *in vitro* cytotoxicity or cytolytic effect of viruses, and also to demonstrate the selectivity or differential susceptibility of viral vectors used in gene therapy [77-86]. An MTS-based cell proliferation assay was qualified earlier for demonstrating IRES mediated selectivity in the cytolytic effect of PVS-RIPO (poliovirus sabine with the internal ribosome entry replaced with its counterpart from the human rhinovirus type 2) [87]. The potential use of *in vitro* cytotoxicity assays as an alternative means for expressing the cytopathic effect of viruses in place of a plaque-forming assay was suggested earlier [87, 88].

2.4.1. MTS-based Cell Proliferation Inhibition Assay and Selectivity of Tropism- Modified Adenoviral Vectors (Ad5-D24-RGD and Ad3/5 D-24)

An MTS-based assay and several fluorescent and chemiluminescence-based cell proliferation or cytotoxicity assays have been used to assess viral cytopathic effects [89-94]. Ad5-D24-RGD is a conditionally replication-competent adenovirus that selectively replicates in tumor cells lacking the Rb (retinoblastoma)/p16 pathway [95]. This adenovirus contains a 24-bp deletion (D24), from bp 923 to 946 of adenovirus type 5 (Ad5), corresponding to the amino acid sequence Leu_{122}-Thr-Cys-His-Glu-Ala-Gly-Phe_{129} of the E1A (adenovirus early antigen) protein known to be necessary for binding the retinoblastoma tumor suppressor/cell cycle regulator protein (Rb). Its fiber knob is modified with an integrin-binding Arg-Gly-Asp (RGD) insertion, allowing for the Coxsackie adenovirus receptor-independent infection of cancer cells [95]. The replication selectivity of Ad5-D-24-RGD is based on the inability of binding the Rb, and, therefore, replication is expected to occur preferentially in cells where S-phase induction is not required, such as in cells defective in the Rb/p16 pathway [95]. This pathway may be faulty in all human cancer cells. The Ad5-D24-RGD virus has been shown to have effective replication and oncolysis in established ovarian cancer cell lines and in a spheroid model using primary ovarian cancer cells purified from malignant ascites as well as in a murine model with peritoneal infected ovarian cancer [96]. This initial investigation suggests that Ad5-D24-RGD may be useful in the treatment of ovarian cancer. Ad5-D24-RGD has been previously shown to kill A549 (a lung carcinoma cell line, ATCC Cat #CCL-

185) cells in a dose-dependent manner, measured by using an XTT ({2,3-bis (2-methoxy-4-nitro-5-sulfophenyl)-5-[(phenylamino) carbonyl]-2H-tetrazolium hydroxide}) cytotoxicity assay [95]. In contrast, MRC5 human embryonic lung fibroblast cell line expresses Rb proteins at normal levels.

Ad5-D24-RGD and Ad3/5–D-24 viral vectors inhibited the proliferation of A549 and MRC5 cells in a dose-dependent manner (Figures 5 A and 5B). The IC_{50} of Ad5-D24-RGD in A549 cells was 10 to 40 fold lower than the IC_{50} in conditions, demonstrating the selectivity of the cell killling effect of Ad5-D24-RGD on A549 and MRC5 cells under identical conditions.The IC_{50} values for the inhibition of A549 cells and MRC5 cells from experiments performed on different days following a standard procedure showed significant inter-day variation.

Although there was variability in the IC_{50} values for both A549 and MRC5 cells, these variations did not show a time or cell passage number-dependent trend. This data suggests that the inclusion of a reference lot in each assay is necessary for lot-lot comparison and stability monitoring, if the assay is to be used as an index of cytolytic activity using a susceptible cell.

For the purpose of demonstrating differences in susceptibility to cytolytic activity, both cell lines must be run within the same plate to avoid assay variations between plates. The variability observed in A549 and MRC5 cells were not uniform, resulting in a higher variation in the differential day-day response on the two cell lines than is usually seen with cell proliferation or proliferation inhibition assays comparing different lots of products like IL-7, IL-15, MR1-1 and HA22 using a single cell line. In seven independent analyses performed using Ad5-D24-RGD over a 19-month period, the ED_{50} values for A549 cells varied from 3.5 to 48.7 virus particles/cell (vp/cell), and the values for MRC5 cells varied from 142.7 to 1078 vp/cell.

The ED_{50} ratio [ED_{50} (MRC5)/ED_{50}(A549)] varied from 33 to 48. The mean ED_{50} value was 34 (CV =65%) for A549 cells; for MRC5 345 (CV=59%). The differential killing effect of Ad5-D24-RGD indicated by the IC_{50} ratio (IC_{50} in MRC5/IC_{50} in A549)] was 27 ± 12 (CV= 45%). The differences in growth characteristics of A549 and MRC5 may have contributed to the high inter-day variations observed. However, at every time point, A549 was at least 10 fold more susceptible to cytolysis than the MRC5 cells.

The MTT-based colorimetric method has been used to quantify cytolytic activity for a number of picornaviruses in cell cultures [97]) and for measuring the infectious titer for a number of parvoviruses [88].

The MTS-based cytotoxicity assay was suggested as an alternative method for defining the infectious titer of the poliovirus in MTS_{50} units under defined conditions analogous to those described earlier for other picornaviruses [87]. The MTS-based cytotoxicity could be qualified for comparing lot-to-lot consistency and stability of the adenoviral vectors by the cytolytic activity of the virus lot in a susceptible cell, such as A549, with a well-characterized reference lot of the viral vector as a standard. The assay needs to be qualified to evaluate the intra-day and inter-day variations, regional effect within the plate, storage effects on stability, and lot-to-lot variations in inter-day assays as described earlier [87]. The viral vector efficacy may be expressed as a function of the dilution that causes the 50% survival as suggested earlier [87]. Both a reference standard and test viral lots need to be tested in the same plate, and the relative IC_{50} values need to be compared.

3. Real-time Dynamic Cell Growth Monitoring

The majority of the cell-based assays are end-point analyses. Real-time cell viability and cytotoxicity monitoring represent the most recent advancements in cell viability assays for continuously monitoring cell growth or death. A number of label-free technologies, including electrical cell-substrate impedance sensing (ECIS), quartz microbalance and optical light mode spectroscopy, and real-time cell electronic sensing (RT-CES), have emerged as a means of monitoring cellular processes in real-time [6, 98-101]. Label-free, noninvasive methods based on electronic cell sensor assays are becoming attractive methods for monitoring cell physiology, particularly adhesion, spreading, and transient changes in cell morphology [101-104]. Real-time cell viability assays, such as those that use the fluorescent reagents like propidium iodide and time-lapse fluorescence imaging [105-106], and_chip-based dielectrophoretic cell immobilization technology for dynamic analysis of drug-induced cytotoxicity [107, 108], are a few examples.

3.1. Real-time Cell Electronic Sensing (RT-CES)

A label-free, non-invasive, real-time electronic sensing (RT-CES) method that continuously measures the attachment and viability of adherent cells was reported recently [98, 109, 110]. In the RT-CES system, microelectronic cell-sensor assays are integrated into the bottom of standard microtiter plates and cells are grown in the individual sensor-containing wells.

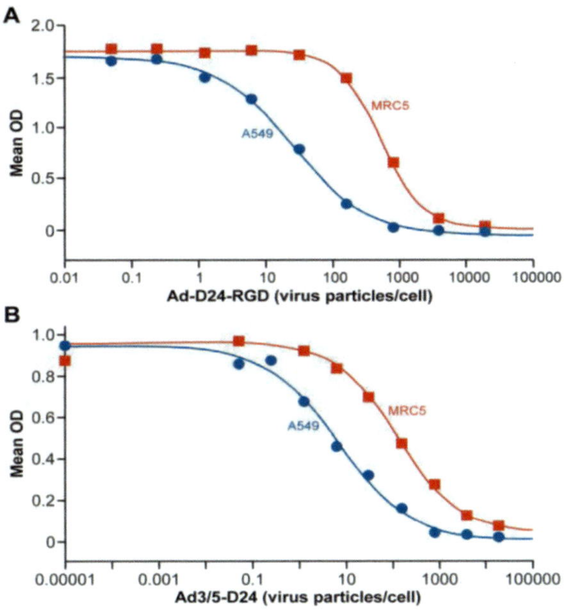

Figure 5. Differential or selective cytolytic effect of conditionally replicating adeno viral vectors. A: A549 cell defective in the Rb gene is more susceptible to cytolysis by Ad5 -D24-RGD compared to the MRC5 cell; B: A549 cell defective in the Rb gene is more susceptible to cytolysis by Ad3/5 -D24 compared to the MRC5 cell.

The electronic sensor provides continuous quantitative information concerning the biological status of the cell impedance. Changes in the biological state of the cell are measured continuously in real-time and recorded. The cell index (CI) is a dimensionless parameter derived as a relative change in measured electrical impedance to represent the cell status. The CI is related to the cell number and is a measure of viable cells attached to the electronic sensor.

Cell adhesion and spreading, receptor-mediated signaling, cell proliferation, compound-, cell- and virus-mediated cytotoxicity, and cellular quality control are a few of the applications cited for the RT-CES system from ACEA Biosciences (San Diego, CA) and Roche known as the xCELLigence system (from Roche). The methodology has the advantage of combining label-free technology with a non-invasive readout with kinetics culminating in information-rich and high-content data that allow users to make well-informed decisions regarding the quality of their assay as well as the quality of the data obtained [111]. RT-CES has been applied to the monitoring of cell adhesion, cell proliferation, and cell death [98, 109, 112-118]. Impedance-based biosensing of the RT-CES system belongs to an emerging technology for analyzing the status of cells *in vitro*. The use of the RTCES system to supplement conventional techniques in pox virology was described recently [119]. The methodology has an application in quantifying virus titers in unknown samples and in developing virus neutralization assays, as well as being a useful tool for the high-throughput characterization of antiviral agents.

3.2. RT-CES Application for Cytoxicity of Immunotoxins

The real-time cell growth of NR6 and NR6M cells (NR6 is a variant of the Swiss 3T3 murine fibroblast cell that does not express EGFR and NR6M is a NR6 cell transfected with EGFRvIII) are measured using RT-CES. The effect of the immunotoxin MR1-1 on the cell growth is shown in Figure 6A.

Continuous real-time monitoring of the cell growth of the two cell lines reveals differences in the growth rate and characteristics of the two cell lines in the culture and the targeted cytotoxicity of the MR1-1 to the EGFRvIII-transfected NR6M cell (a result that confirms earlier data from a [^{3}H] lecucine incorporation assay [70] and MTS assay described in section 2.3.1 of this chapter). NR6 cells showed a continuous growth for four days, after which they reached a confluent stage and started to decline in growth after five days (cell death). However, NR6M cells exhibited continuous growth for six days under experimental conditions. NR6M cells showed a significantly lower CI with 0.2 ng/ml of MR1-1 and were completely killed within 24 hours of receiving an addition of 2 ng/ml MR1-1. MR1-1 had no effect on the control cell line NR6. The immunotoxin HA22 (targeting CD22) had no effect on the NR6M cell proliferation. The cytoxicity of MR1-1 on NR6M cell was dose-dependent. The results from MTS colorimetric and RT-CES assays on the target specific cytotoxicity were comparable (Figure 3A. 6A, 6B, and 6C). Whereas the [^{3}H] leucine and MTS assays were end-point assays showing the cumulative effect on the cell growth for a certain incubation period, the RT-CES continuously monitored the effect in real time from the beginning to the end of the experimental period.

3.3. RT-CES and Selective Cytolysis with Tropism-modified Adenoviral Vectors

In vitro selective cytolysis by viral vectors is currently receiving more attention as a means for documenting the target cell selectivity before entering safety and efficacy studies with human subjects. We have demonstrated earlier the MTS-based cell proliferation assay for documenting selective or differential cytolytic activity in cells. Ad5 –D24-RGD and Ad3/5-D24 were tested for differential cytolytic response to cell lines with differing expression of Rb protein. A549 cells, which are deficient in Rb, were more susceptible to cytolysis by both adeno viral vectors. The cytolytic efficiency was highly dependent on the length of the culture after the addition of the viral vectors. Real-time, continuous monitoring of the viral effect would provide more insight into the cell growth stage.

The cell growth kinetics of A549 and MRC-5 cells at different cell seed densities in the RT-CES electrode-impeded wells of a multi-well plate are shown in Figure 7A. The CI values are higher for A549 cells at all seed densities for the entire duration of the culture, suggesting that A549 grows at a faster rate. Figure 7B shows the growth rate kinetics of the A549 cells at different cell seed densities. At low seed density (1,000 cells/well) both cells are in a slow growth phase for 2-to-3 days (Figure 7C). The MRC5 cell index is lower than that of A549 during the first 3 days post-seed, The CI value for MRC5 increases at faster rate compared to that of A549 during days 4-6. After day 6, A549 is still in the growth phase with the CI values increasing whereas the MRC5 reaches stationary phase. The linear growth curve-fit and the slope of the linear curve-fits at different time periods (Table 4) provides a good understanding of the differential cell growth kinetics of the two cell lines.

Table 4. A549 and MRC5 cells show differences in growth rate characteristics monitored by RT-CES

Post-seed culture time	Growth rate (slope of linear growth plot) (CI vs time)		Linear curve fit coefficient of correlation (R^2)	
Hours	A549	MRC5	A549	MRC5
0-50	0.0058	0.0016	0.993	0.806
0-80	0.0077	0.0041	0.916	0.768
70-150	0.0465	0.057	0.98	0.989
70-170	0.048	0.049	0.989	0.966

The cytolysis of A549 and MRC5 cells by Ad5-D24-RGD was measured using the RT-CES method is shown in Figure 8A and 8B. There was much higher cytolysis (detachment of cells) of A549 cells compared to MRC 5 cells. With A549 cell, cytolysis was evident after 75 hours with 20 vp/cell and 45 hours with 200 vp/cell. With MRC5, cell cytolysis did not occurr at 100 hours with a viral vector dose of 200 vp/cell, and the cells showed no effect up to 130 hours with 2 or 20 vp/cell. RT-CES shows a time- and viral infection dose- dependent effect of cytolytic activity with both cell lines in a single experiment. A comparison of the IC_{50} values of the cytolysis of A549 and MRC5 cells by Ad5–D-24-RGD at different time points after the addition of viral vector showed that at all time periods A549 cells are more susceptible to cytolysis than MRC5 cells.

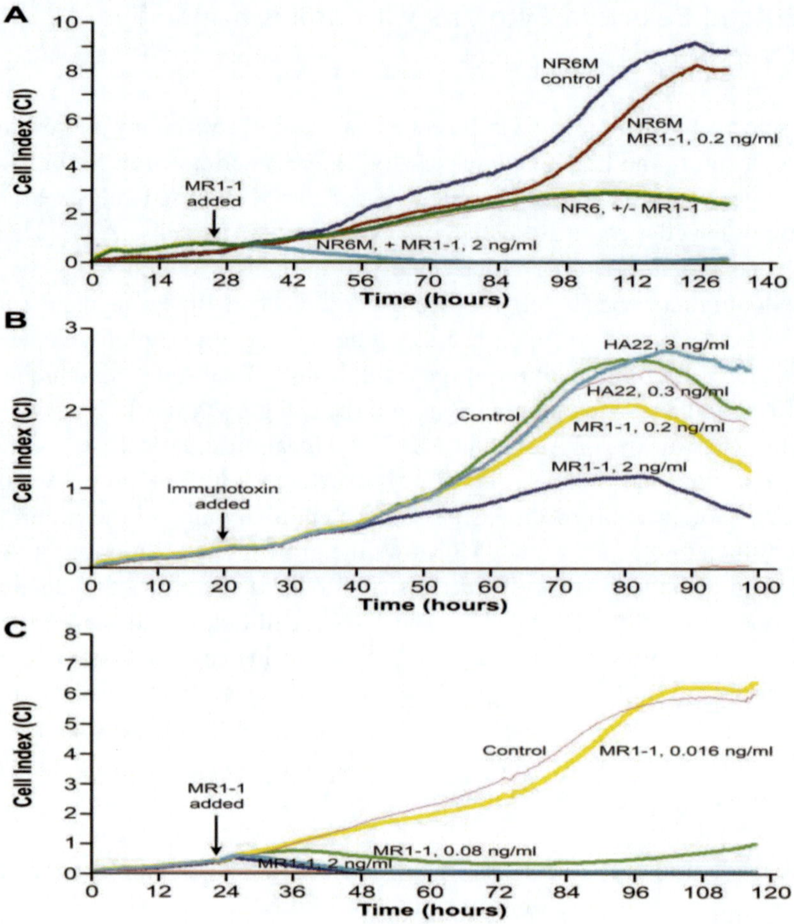

Figure 6. Selectivity of cytotoxicity of MR1-1 to the EGFRvIII gene transfected NR6M cell as probed by the RT-CES. A: MR1-1 is selective to NR6M compared to control parent cell line NR6; B: MR1-1 is cytotoxic to EGFRvIII-expressing cells but HA22 is not; C: Concentration dependence of cytotoxicity of MR1-1 on NR6M cells.

The conclusions from the RT-CES analysis were consistent with the observation from the MTS cell proliferation assay. The inter-day variation in IC_{50} values for A549 cells and MRC5 cells, as well as the relative cytolytic effect, may be explained by the difference in the relative growth rate and the viral sensitivity.

The absolute ED_{50} values of the cytolytic effect on either cell line as measured by the MTS and RT-CES methods cannot be directly compared because of the different conditions of assay monitoring set ups and the measurement of different cell characteristics. The RT-CES method requires a lower number of cells, and the whole experiment could be performed on the same plate with no additional reagents. The cytolytic activity was dependent on the cell growth stage of the cells in both assays. There was very little cell killing during day 1 and day 2 post-infection, and the ED_{50}values decreased with increasing time post-infection, reaching nearly a complete killing of the cells within 5-6 days post-infection.

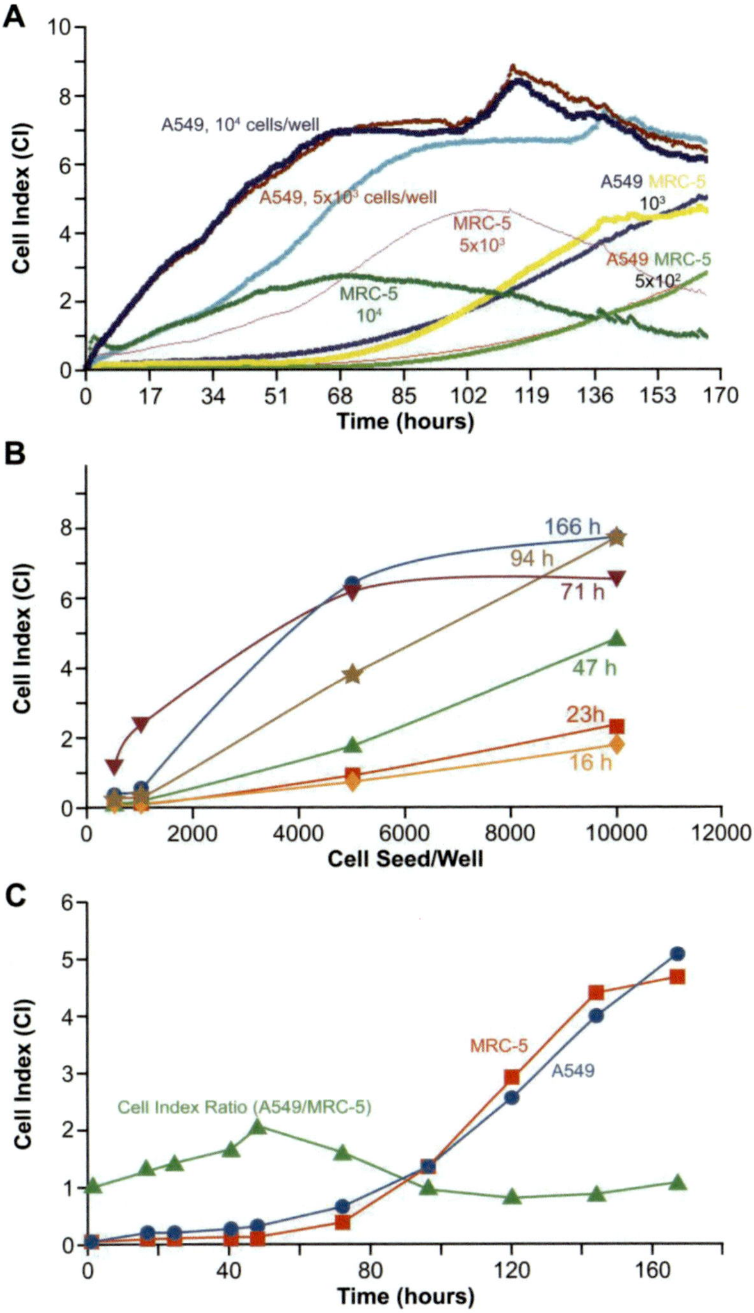

Figure 7. Real-time dynamic monitoring of the growth rates of A549 and MRC5 cells with the RT-CES system. A: Growth curves at different cell seed densities; B: Cell Index (Cell number) as a function of cell seed density at different time periods; C: A549 and MRC5 growth curves show differences in growth kinetic.

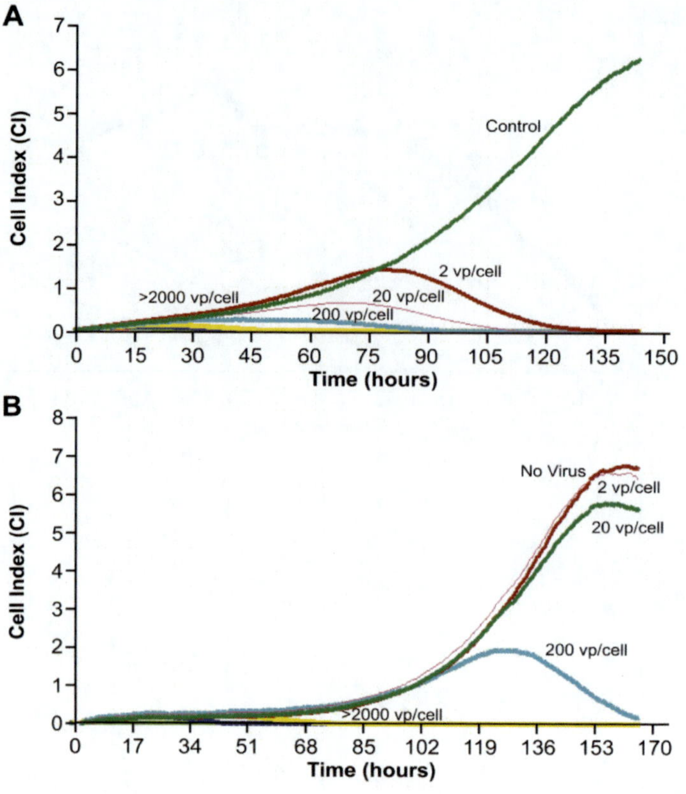

Figure 8. Cytolysis of A549 cells (A) and MRC5 cells (B) with Ad5-D24-RGD probed using RT-CES; vp/cell: virus particles/cell.

4. Signal Transduction and Cell Proliferation (Reporter Gene Assays)

Simple growth and viability assays have the advantages of convenience, but do not immediately reveal insights into the action mechanisms of compounds. Recombinant technology has opened the door for mechanistic cell-based assays by engineering cells to express particular molecular targets. A large number of transfected or recombinant cell lines with different reporter genes are currently available. Using these recombinant cell lines with reporter genes has improved the speed of research and development in the areas of protein production as well as mechanism-based cell-based assays [8, 120, 121]. The reporter gene system is a technology in which a reporter gene is synthesized in response to the activation of a specific signaling cascade of interest, followed by monitoring the reporter protein expression through its enzymatic activity linked with a variety of colorimetric, fluorescent or luminescent read-outs. Due to its inherent sensitivity, large signal dynamics and simplicity of set up, reporter assay platforms have been widely used in high-throughput homogenous assays for screening.

4.1. STAT3 Luciferase Reporter Gene Assay

The signal transducer and activator of transcription 3 (STAT3) was originally termed the acute-phase response factor (APRF) because it was identified as a DNA-binding activity within IL-6-stimulated hepatocytes, and was capable of selectively interacting with an enhancer element in the promoter complex of acute-phase response genes [122]. APRF was subsequently shown to be one of the seven known signal transducers and activators of transcription (STAT) proteins involved in cell proliferation, differentiation, and apoptosis. STAT3 controls a key signaling pathway in the development of many malignant diseases. Several genetic studies have proven its central role in regulating proliferation, apoptosis, angiogenesis and immune responses, making it an attractive target for cancer therapy [123]. STAT3 acts mainly as a signal transducer of the IL-6 family of cytokines and transcriptionally activates specific target genes. It is the main member of the STAT family, activated by the IL-6 family of cytokines; and it has been recognized as an essential molecule for gp130-mediated cell growth, survival, and differentiation. It also functions as an oncogene and is required for transformation, enhancing transformation, or blocking apoptosis.

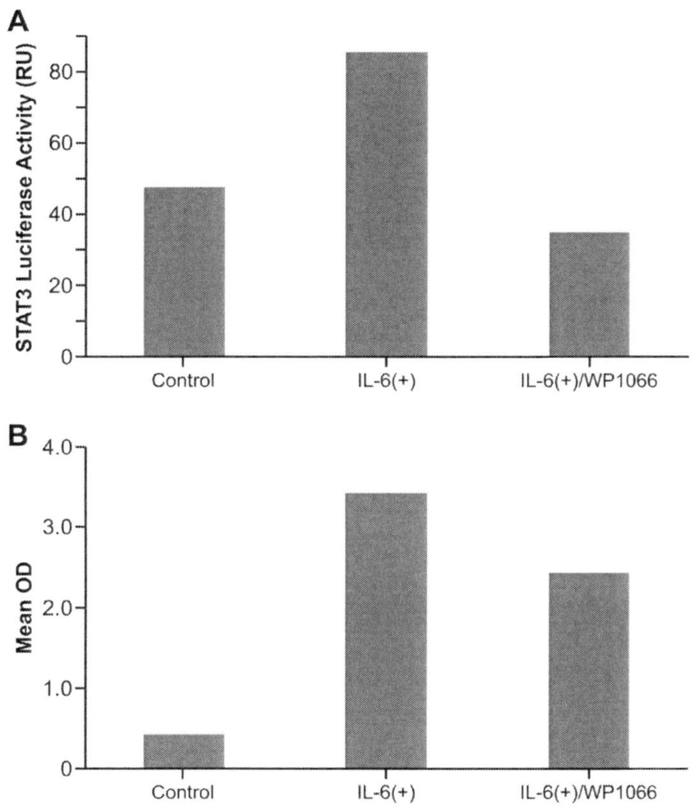

Figure 9. STAT3 reporter gene assay and cell proliferation. A: Effect of STAT3 modulators on reporter gene expression in Hela STAT3-Luc cells determined by luciferase activity measurement; B: Effect of STAT3 modulators on the proliferation Hela STAT3-Luc cells determined by an MTS cell proliferation assay.

Oncostatin M (OSM) is a member of the IL-6 cytokine family, produced by inflammatory cells and some tumor cells. Following OSM binding to its receptor and glycoprotein (gp130), JAK2 (Janus kinase 2) is phosphorylated, which in turn phosphorylates STAT3, permitting nuclear translocation and modulation of gene expression. Several transcriptional targets of STAT3 are important contributors to tumor biology and the activation of STAT3 by gp130-mediated mechanisms, known to be oncogenic. STAT3 has been identified as an important target for cancer therapy since it participates in oncogenesis through the up-regulation of genes, encoding cell cycle regulators (cyclin D1/2 and c-myc), apoptosis inhibitors (Bcl-xL, Bcl-2, Mcl-1, and survivine), and inducers of angiogenesis (VEGF). STAT3 is also known to be constitutively active in patients with cancers including prostate cancers, breast cancers, and lung cancers and in more than 90% of head and neck cancers [122, 124].

A number of STAT3 reporter gene cell lines have been constructed and utilized for studying the role of STAT3 in regulating cellular activities. A STAT3 Luc DU-145 prostate cancer cell line was constructed and regarded as a rapid, sensitive, and cost-effective method for screening potential STAT3 modulators [125]. A stable Hela STAT3-Luc cell is available from Panomics/ Affymetrix (Santa Clara, CA).

The STAT3-Luc expression, as monitored by luciferase activity (Figure 9A), showed a low basal expression of STAT3 in the absence of a STAT3 stimulator, enhanced STAT3-Luc expression in the presence of the STAT3 stimulator IL-6 and inhibition of STAT3-Luc by the JAK3 kinase inhibitor WP1066 {(E)-3(6-bromopyridin-2-yl)-2-cyano-N-(S0-1phenylethyl) acrylamide}, a reported STAT3 inhibitor, consistent with results demonstrated by the provider of the cell line [Panomics/Affymetrix (Santa Clara, CA)].

Reporter assays have to be validated by showing that the compounds tested are not interfering with the reporter system itself. They also need to be tested directly against the presumed target to show that they are acting through the desired mechanism. For example, a screen of marine natural products using a cell line with a luciferase reporter linked to the Wnt {a hybrid of IntI (integration1) and Wg (wingless)- a network of proteins} pathway revealed compounds that indirectly affected the pathway through inhibiting histone deacetylase (HDAC) [126]. The STAT3-Luc activity regulation by the modulator IL-6 and WP1066 correlated with the cell proliferation activity monitored with the MTS cell proliferation assay (Figure 9B).

The stimulation of the STAT3-Luc activity by the positive modulator OSM was OSM-dose dependent (Figure 10A). Similarly the inhibition of the OSM stimulated STAT3-Luc activity by the inhibitor WP1066, dependent on the inhibitor concentration (Figure 10B). These results are consistent with the results reported from earlier studies from Panomics/Affymetrix (Santa Clara, CA). Once the reporter gene assay is validated, the reporter cells provide a useful tool for following gene or protein expression evaluation and establishing mechanism-based, cell-based assays for drug discovery and development efforts. A prostate cell DU-145-STAT3-Luc expression system was used earlier to evaluate the STAT3 modulation of prostate cancer [125]. As with any other cell-based assay, the reporter gene-based assay is also subject to artifacts due to the drug molecule itself or impurities in drug molecule preparations, as well as to components of the drug formulation or drug delivery vehicle.

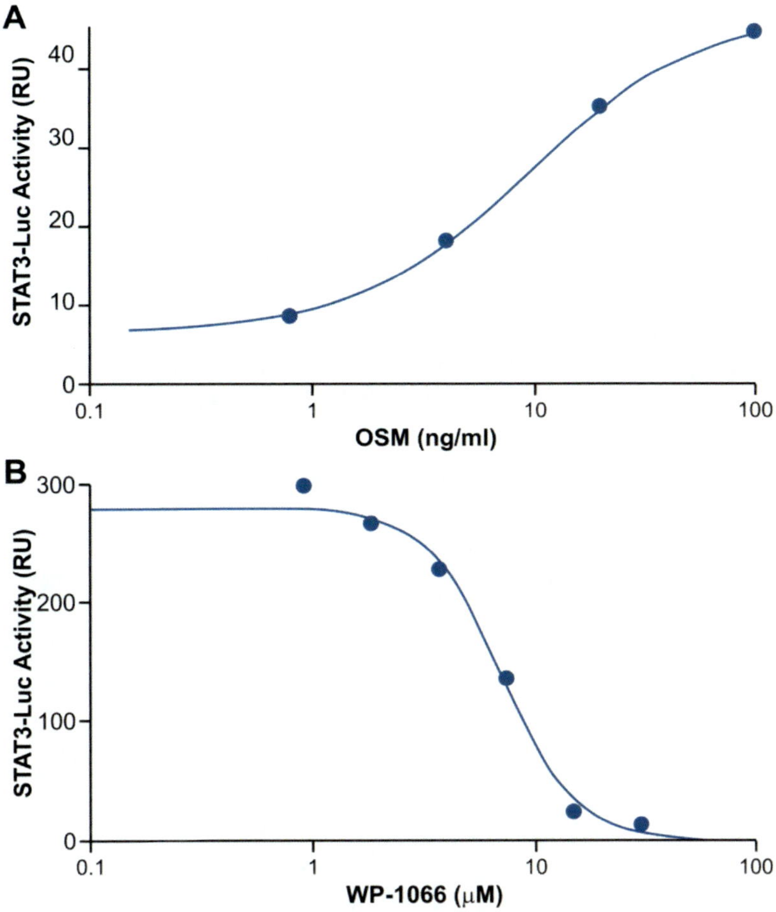

Figure 10. Modulation of STAT3 activity. A: Oncostatin M dose dependent stimulation of STAT3-Luc gene expression measured using Luciferase activity assay; B: WP1066 dose-dependent inhibition of STAT3-Luc activity in a HeLa STAT-3 Luc cell.

Conclusion

The non-radioactive, MTS dye-based colorimetric cell proliferation assay provides a simple, cost-effective assay for monitoring the proliferation of cells in the presence and absence of any stimulatory or inhibitory factors. This assay can be easily qualified and validated as *in vitro* cell-based assays, for monitoring the effect of growth factors such as cytokines, interferon etc., and the cytotoxicity of molecules, such as immunotoxins, on susceptible cells to establish product quality for controlled studies such as clinical investigations. The assay can be used to monitor the cytotoxic effect of molecules by following the cell proliferation inhibition effect.

The use of the cell proliferation assay to monitor the biological activity of growth factors is demonstrated using IL-7. The cell proliferation inhibition assay and its applications in targeted or selective cytotoxicity are demonstrated using the immunotoxins MR1-1 and HA22.

The selective cytolysis of replication-deficient and conditionally replicating viral vectors has been investigated. The MTS-based proliferation assay and the dynamic electrode impedance cell attachment assay using RT-CES are compared for use in cytotoxicity and selective cytolytic activity of immunotoxins and viral vectors.

The assay conditions for testing drug stability during clinical administration are demonstrated, as is monitoring functionally active drug molecules in body fluids.

The reporter gene assays, once validated, provide a mechanism-based, cell-based assay to assist in drug development efforts. This provides a rapid and cost effective analytical tool for drug development as well as clinical evaluation. The combination of conventional cell proliferation, cell death and cytopathic assays are currently being utilized in the drug discovery stage to gain a better understanding of the mechanistic aspects of drug action.

The xCELLigence System, based on the RT-CES technology and other real-time dynamic monitoring systems, allows for the label-free and real-time monitoring of cellular processes, such as cell proliferation, cytotoxicity, adhesion, viability, invasion, and migration, using electronic cell-sensor array technology. Real-time monitoring of cellular processes by the xCELLigence System offers distinct and important advantages over traditional end-point assays. First, the avoidance of labels allows for more physiologically relevant assays, saving on time, labor, and resources. Second, a comprehensive representation of the entire length of the assay is possible, enabling the user to make informed decisions regarding the timing of certain manipulations or treatments. Finally, the actual kinetic response of the cells within an assay prior or subsequent to certain manipulations provides important information regarding the biological status of the cell such as cell growth, arrest, morphological changes, and apoptosis.

Acknowledgments

This work has been funded in whole or in part with federal funds from the National Cancer Institute, National Institutes of Health, under Contract No. HHSN261200800001E. The content of this publication does not necessarily reflect the views or policies of the Department of Health and Human Services, nor does mention of trade names, commercial products, or organizations imply endorsement by the U.S. government. This research was supported (in part) by the Developmental Therapeutics Program in the Division of Cancer Treatment and Diagnosis of the National Cancer Institute.

We would like to acknowledge the support and encouragement from Dr. Stephen P. Creekmore (chief of Biological Resources Branch of NCI). We would also like to acknowledge the technical support from Ms. Nirmala Saptharshi and Mr. Hengguang Jiang. The critical review of this chapter by Mr. Trevor Broadt of BDP is greatly appreciated. The generous supply of the products, from the Biopharmaceutical Development Program (BDP; SAIC-Frederick, Inc.) manufacturing group, used in this study is greatly appreciated. We also would like to thank the BDP process analytics group for providing the infusion mock model processed samples.

References

[1] D. Verduzco, J. F. Amatruda, *Methods in Cell Biology,* 101, 19 (2011).

[2] L. Galluzzi, S. A. Aaronson, J. Abrams, E. S. Alnemri, D. W. Andrews and E. H. Baehrecke et al., *Cell Death and Differentiation* 16, 1093 (2009).

[3] M. J. Stoddart, P. I. Furlong, A. Simpson, C. M. Davies and R. G. Richards, *European Cells & Materials*, 12, 16. (2006).

[4] M. J. Stoddart, (ed), Mammalian Cell viability, Methods and protocols. Methods Mol Biol, 740, (2011).

[5] M. J. Stoddart, In Stoddart, MJ. (ed), Mammalian Cell viability. Methods and protocols. Methods, Mol Biol.2011, 740, 1 (2011).

[6] L. K. Minor, *Combinatorial Chemistry & High-Throughput Screening*, 11, 573 (2008).

[7] J. Hong, K. Kandasamy, M. Marimuthu, C. Soo Choi, and S. Kim, *Analyst*, 136, 237(2011).

[8] A. L. Harvey and I. A. Cree, *Planta Medica*, 76, 1080 (2010).

[9] R. H. Shoemaker, *Nature Reviews Cancer*) 6, 813 (2006).

[10] K. IljinK, P. Ketola, P. Vainio, Halonen, P. Kohonen, V. Fey, R. C, Grafströmm, M. Perälä and O. Kallioniemi, *Clinical Cancer Research*, 15, 6070, (2009).

[11] E. A. Guzman, J. D. Johnson, M. K. Carrier, C. I. Meyer, T. P. Pitts, S. P. Gunasekera and A. E. Wright, *Anticancer Drugs*, 20, 149 (2009).

[12] O. Kepp, L. Galluzzi, M. Lipinski, J. Yuan and G. Kroemer, *Nature Reviews Drug Discovery*10, 221 *(2011).*

[13] K. Berg, L. Zhai, M. Chen, A. Kharazmi and T, C. Owen, *Parasitology Research*, 80, 235(1994).

[14] G. Malich, B. Markovic and C. Winder, *Toxicology*, 124, 179 (1997).

[15] G. Soman, X. Y Yang, J. Hengguang, S. Giardina, V. Vyas, G. Mitra, J. Yovandich, S. P. Creekmore, T.A. Waldmann and W. G. Alvord, *Journal of Immunological Methods* 348, 83 (2009).

[16] J. R. Foster, *International Journal of Experimental Pathology*, 82, 171 (2001).

[17] S. R. Hofmann, R. Ettinger, Y. J. Zhou, M. Gadina, P. Lipsky, R. Siegel, F. Candotti and J. J. O'Shea, *Current Opinion in Allergy and Clinical Immunology,* 2, 495 (2002).

[18] T. Mossman, *Journal of Immunological Methods*, 65, 55 (1983).

[19] A. Lalvani, P. L. Meroni, K. A. Millington, M. L.Modolo, M. Plebani, A. Tincani, D. Villalta, A. Doria and A. Ghirardello, *Clinical and Experimental Rheumatology*, 26 (1 Suppl 48), S62(2008).

[20] K. S. Lee, J.H. Chung, T.K Toy, S. Y. Suh, B.H. Oh and C.H. Hong, *Dementia and Geriatric Cognitive Disorders,* 28, 281, (2009).

[21] N. Sachdeva and D. Asthana, *Frontiers in Bioscience*, 12, 4682(2007).

[22] X. Zhou, M. S, Fragala, J. E. McElhaney and G.A. Kuchel, *Current Opinion in Clinical Nutrition Metabolic Care*, 13, 541 (2010).

[23] S. Roy-Ghanta and J. S. Orange, *Clinical Reviews in Allergy and Immunology* 38, 39 (2010).

[24] A. R Mire-Sluis, L Page, and R. Thorpe, *Journal of Immunological Methods*, 187, 191(1995).

[25] A. Meager, Methods, 2005 38, 237.

[26] A. M. Crawley and J. B. Angel, *Immunology and Cell Biology*, 2011 doi: 10.1038/icb.2011.66.

[27] W. Liao, J. X. Lin and W. J. Leonard, *Current Opinion in Immunology*, 23, 598 (2011).

[28] A. E. Namen, S. Lupton, K. Hjerrild, J. Wignall, D.Y. Mochizuki, A. Schmierer, *et al., Nature*, 333, 571 (1988).

[29] K.S. Schluns, W.C. Kieper, S. C. Jameson and L. Lefrancois, *Nature Immunology,*1, 426, (2000).

[30] C. D. Surh, O. Boyman, J. F. Purton and J. Sprent, *Immunol Rev*. 211, 154 (2006).

[31] J. T. Tan, E. Dudl, E. LeRoy, R. Murray, J. Sprent K. I. Weinberg and C.D. Surh, *Proceedings of the National Academy of Sciences of the United States of America,* 98, 8732 (2001).

[32] L, Willems, S. Li, O. Rutgeerts, C. Lenaerts, M. Waer and A. D. Billiau, *Journal of Immunology*, 187, 3587 (2011).

[33] J. Ming, G. Jiang, Q. Zhang, X. Qiu, and E. Wang, *Cancer Immunology, Immunotherapy,* 2011 Aug 17. [Epub ahead of print]

[34] R. Li, A. Paul, K. W. Ko, M. Sheldon, B. E. Rich, T. Terashima, C. Dieker, S. Cormier, L Li, E. A. Nour, L. Chan,K. Oka, *European Heart Journal*, 2011 Jul 30, [Epub ahead of print]

[35] P. Tsapogas, S. Zandi, J. Ahsberg, J. Zetterblad, W. Eelinder J. I.Jönsson, R. Månsson, H. Qian, and M. Sigvardsson, *Blood*, 118, 1283 (2011).

[36] A. Silva, B. Laranjeira, L. R Martins, B. A. Cardoso, J. Demengeot, J. A. Yunes, B. Seddon and J. T. Barata, *Cancer Research*, 71, 4780 (2011).

[37] S. D. Katzman, K. K. Hoyer, H. Dooms, I. K Gratz, M, D. Rosenblum, J. S. Paw, S. H. Isakson, and A. K. Abbas, *Cytokine,* 56, 116 (2011).

[38] C. L. Mackall, T. J. Fry, and R. E Gress, *Nature Reviews: Immunology, 11*, 330 (2011).

[39] H. E. Fleming and C. J. Paige, Seminars in Immunology, 14, 423 (2002).

[40] K. Ishihara, K. Medina, S-I. Hayashi, C. Pietrangeli, A. E. Namen, K. Miyake, and P. W. Kincade, *Dev Immunol.*, 1,149 (1991).

[41] G. Lee, A. E. Namen, S. Gillis and P. W. Kincade, *Current Topics Microbiology and Immunology*, 141, 16 (1988).

[42] G. Lee, K. Medinaand and P. W. Kincade, *Current Topics Microbiology and* Immunology, 152, 33 (1989).

[43] G. Lee, A. E. Namen, S. Gillis, L. R. Ellingsworth, and P. W. Kincade, *Journal of Immunology,* 142, 3875 (1989).

[44] A. R. Mire-Sluis, L. Healey, S. Griffiths, D. Hockley, and R Thorpe, *J. Immunol.Methods*, 236, 71 (2000).

[45] L. S. Park, D. J. Friend, A. E. Schmierer, S.K. Dower, and A. E. Namen, *J. Exp. Med.*171, 1073 (1990).

[46] C. E. Pietrangeli, S. Hayashi, and P. W. Kincade, *Eur. J. Immunol.,*18, 863, *(*1988).

[47] C. D. Milne, H. E. Fleming, Y. Zhang and C. J. Paige, *Developmental Immunology*, 1, 149{1991).

[48] R. G. Goodwin, S. Lupton, A. Schmiere, K. J. Hjerrild, R. Jerzy, W. Clvenger, S. Gillis, D. Cosman and A. E. Namen, *Proceedings of the National Academy of Sciences of the United States of America*, 86, 302 (1989).

[49] H. R. Kim, K. A. Hwang, S. H. Park and I. Kang , *Critical Reviews in Immunology*, 28, 325 (2008).

[50] A. Khatri, Y. Husaini and P. J. Russell, *Journal of Immunological Methods*, 326, 41 (2007).

[51] L. Lam, C. Lam and Y. Cao, *Drugs of the Future,* 29, 609 (2004).

152] R. J. Kreitman, *American Association of Pharmaceutical Scientists journal*, 8, E532 (2006).

[53] I. Pastan, and R. J. Kreitman, *Advanced Drug Delivery Reviews*, 31, 53 (1998).

[54] A. Boskovitz, C. J. Wikstrand, C. T. Kuan, M. R. Zalutsky, D.A. Reardon, and D. D. Bigner, *Expert Opinion on Biological Therapy,* 4, (2004).

[55] M. F. Chiron, *Bulletin du Cancer,* 84, 1135 (1997).

[56] S. Choudhary, M. Mathew and R. S. Verma, *Drug Discovery Today,* 16, 495 (2011).

[57] D. J. R. FitzGerald, W. Kreitman, D. Wilson Squires and I. Pastan, *Journal of Medical Microbiology*, 293, 577 (2004).

[58] F. J. Hernandez-Ilizaliturri, A. Gowda and M. S. Czucz, *Clinical Advances in Hematology & Oncology*, 2, 606 (2004).

[59] B. Hirsch, J. Brauer, M. Fischdick, C. Loddenkemper, S. Bulfone-Paus, H. Stein, and H. Dürkop, *Current Drug Targets*, 10, 110 (2009).

[60] K. Kawakami, B. J. Aggarwal, and R. K. Puri, (Eds). Cytotoxins and Immunotoxins for Cancer Therapy. CRC Press (2004).

[61] R. J. Kreitman, *Current Pharmaceutical Design*, 15, 2652 (2009).

[62] R. J. Kreitman, *BioDrugs*, 23, 1 (2009).

[63] R. J. Kreitman, E. Arons, M. Stetler-Stevenson, D. J. Fitzgerald, W. H. Wilson, and I. Pastan, *Leukemia & Lymphoma*, 52 Suppl 2, 82 (2011).

[64] I. Pastan, R. Hassan, D. J. FitzGerald and R. J. Kreitman, *Nature Reviews:Cancer,* 6, 559 *(2006).*

[65] S. Potala, S. K. Sahoo and R. S. Verma, *Drug Discovery Today*, 13, 807(2008).

[66] S. Shapira, V. Lisiansky, N. Arber, and S. Kraus, *Expert Opinion on Investigational Drugs*, Suppl 1, S67 (2010).

[67] M. K. Tur, I. Neef, G. Jäger, A. Teubner, M. Stöcke, G. Melmer, and S. Barth, *Current Pharmaceutical Design*, 15, 2693 (2009).

[68] J. K. Vasir and V. Labhasetwar, *Technology in Cancer Research & Treatment*, 4, 363 (2005).

[69] S. Bang, S. Nagaria, M. Onda, R. J. Kreitman, and I. Pastan, *Clinical Cancer Research, 11,* 1545 (2005).

[70] I. A. J. Lorimer, A. Keppler-Hafkemeyer, R. A. Beers, C. N. Pegram, D. D. Bigner and I. Pastan, *Proceedings of the National Academy of Sciences of the United States of America*, 93, 14815 (1996).

[71] R. J. Kreitman, I. Margulies, M. Stetler-Stevenson, Q-C. Wang, D. J. FitzGerald, J. P. David and I. Pastan, *Clinical Cancer Research, 6,* 1476 (2000).

[72] E. Mansfield, P. Amlot, I. Pastan, and D. J. FitzGerald, *Blood*, 90, 2020 (1997).

[73] C. T. Kuan, C. J. Wikstrand, and D. D. Bigner, *Brain Tumor Patholology*, 17, 71 (2000).

[74] C. T. Kuan, C. J. Wikstrand, G. Archer, R. Beers, I. Pastan, M. R. Zalutsky and D. D. Bigner, *International Journal of Cancer*, 88, 962 (2000).

[75] H. Ochiai, G. E. Archer, J. E. 2nd.Herndon, C. T. Kuan, D. A. Mitchell, D. D. Bigner, I. Pastan and J. H. Sampson, *Cancer Immunology, Immunotherapy*, 857,115 (2008).
[76] S. A. Vorberg, and K. K. Hunt, *Oncologist*, 7, 46 (2002).
[77] R. E. Adamson, A. A. Frazier, H. Evans, K. F.Chambers, E. Schenk, M. Essand, R. Birnie, R. R. Mitry, A. Dhawanm and N. J. Maitl, *Human gene therapy*, Epub ahead of print, DOI: 10.1089/hum.2011.021.
[78] M. H. Collins, H. Chong, C. H. Ahn, T. P. Cripe, PLoS ONE | www.plosone.org 1, 6, | Issue 5 | e19530 (2011).
[79] A. Hemminki, A. Kanerva, E. J. Kremer, G. J Bauerschmitz, B. F. Smith, B. Liu, M. Wang, R. A. Desmond, A. Keriel, B. Barnett, H. I. Baker, G. P. Siegal and D. T. Curiel, *Molecular Therapy,* 7, 163 (2003).
[80] L. Heinemann, G. R. Simpson, A. Boxall, T. Kottke, K. L. Relph, R. Vile, A. Melcher, R. Prestwich, K. J. Harrington, R. Morgan, and H. S. Pandha, *BMC Cancer*, 11, 221(2011).
[81] J. L. Jakubczak, P. Ryan, M. Gorziglia, L. Clarke, L. K. Hawkins, C. Hay, Y. Huang, M. Kaloss, A. Marinov, S. Phipps, A. Pinkstaff, P. Shirley, Y. Skriphenko and D. Stewart, *Cancer Research*, *63*, 1490 (2003).
[82] K Yamada, H. Moriyama, H. Yasuda, K. Hara, Y. Maniwa, H. Hamada, K. Yokono and M. Nagata, *Human Gene Therapy, 18*, 389 (2007).
[83] C. Gomez-Manzano, W. K. Yung, R. Alemany and J. Fueyo, *Neurology*, 63, 418 (2004).
[84] R. P. Rocconi, M. T. Numnum, Z. B. Zhu, B. Lu, M. Wang, A. A. Rivera, M. Stoff-Khalili, R. D. Alvarez, D. T. Curiel and S. Makhija, *Gynecologic Oncology,* 103, 985 (2006).
[85] Z.B. Zhu, B. Lu, M. Park, S. K. Makhija, T. M. Numnum, J. E. Kendrick, M Wang, Y. Tsuruta, P. Fisher, R. D. Alvarez, F. Zhou, G. P. Siegal, H. Wu and D. T. Curiel, *International Journal of Oncology,* 32, 1179 (2008).
[86] Y. Liu, T. Ye, D. Sun, J. Maynard and A.Deisseroth, *Human Gene Therapy*, 15, 637 (2004).
[87] X. Y.Yang, E. Chen, H. Jiang, K. Muszynski, R. H. Harris, S. Giardina, G. Mitra, M. Gromeier and G. Soman, Journal of Virological Methods, 155, 44 (2009).
[88] C. L. Heldt, R. Hernandez, U. Mudiganti, P.V. Gurgel, D. T. Brown and R. G. Carbonell, *Journal of Virological Methods*, 135, 56 (2006).
[89] A. Malecki, R. Garrido, M. P. Mattson, B. Henning, and M. Toborek, *Journal of Neurochemistry*, 74, 2278 (2000).
[90] K. Heese, Y. Nagai and T. Swada, *Neuroscience Letters*, 288, 37 (2000).
[91] J. Gauduchon, F. Gouilleux, S. Maillardm, V. Marsaud, J. Renoir and B. Sola, *Clinical Cancer Research*, 11, 2345 (2005).
[92] C. Zhang, P. Hazarika, X. Ni, D. Weidner and M. Duvic, *Clinical Cancer Research*, 8, 1234 (2002).
[93] P. Giridharan, S. T. Somamsundaram, K. Perumal, R. A. Vishwakarma, N. P. Karthikeyan, R. Velmurugan and A. Balakrishnan, *British Journal of Cancer*, 87, 98 (2002).
[94] F. Tonello, M. Seweso, O. Marin, M. Mock and C. Montecuccp, Nature, 418, 386 (2002).

[95] K. Suzuki, J. Juan Fueyo, V. Krasnykh, P. N. Reynolds, D.T. Curiel and R. A. Alemany *Clin. Cancer Res.* 7,120 (2001).

[96] G. J. Bauerschmitz, J. T. Lam, A. Kanerva, K. Suzuki, O. M. Nettelbeck, J. Dmitriev, V. Krasnykh, G.V. Mikheeva, M. N. Barnes, R. C. Alvarez, P. Dall, R. Alemany, D. T. Curiel, and A. Hemminki, *Cancer Research*, 62, 1266 (2002).

[97] P. Andersson, S. Alm, K. Edman and A. M. Lindberg, *Journal of Virological Methods*, 130, 117 (2005).

[98] J. M. Atienza, N. Yu, S. L. Kirstein, B. Xi, X. Wang, X. Xu and Y. A. Abassi, *Assay and Drug Development Technologies*, 4, 597 (2006).

[99] T. S. Hug, *Assay and Drug Development Technologies*, 1, 479 (2003).

[100] J. Voros, R. Graf, G. L. Kenausis, A. Bruinink, J. Mayer, M.Textor, E. Wintermantel and N. D. Spencer, *Biosensors & Bioelectronics*, 15, 423 (2000).

[101] B. Xi, N. Yu, X. Wang, X. Xu and Y. Abassi, *Biotechnology Journal*, 3, 484 (2008).

[102] A. M. Otto, M. Brischwein, H. Grothe, E. Motrescu and B. Wolf, Recent Results Cancer Res. 161, 39 (2003).

[103] C. R. Keese, K. Bhawe, J. Wegener, I, Giaever, *Biotechniques*, 33, 842 (2002).

[104] C. Spegel, A. Heiskanen, L. L. Henrik D. Skjolding and J. Emnéus, *Electroanalysis*, 20, 680 (2008).

[105] D. Wlodkowic, S. Faley, Z. Darzynkiewicz and J. M. Cooper, *Methods in Molecular Biology,* 731, 259 (2011).

[106] H. Zhao, J. Oczos, P. Janowski, D. Trembecka, J. Dobrucki, Z. Darzynkiewicz and D. Wlodkowic, *Cytometry Part A*, 77, 399 (2010).

[107] K. Khoshmanesh, J. Akagi, S. Nahavandi, J. Skommer, S. Baratchi, J. M. Cooper, K. Kalantar-Zadeh, D. E. Williams and D. Wlodkowic , Anal Chem. 83, 2133 (2011).

[108] H. Hatanaka,T. Yasukawa and F.Mizutani, Anal Chem. *83, 7207(*2011).

[109] S. L. Kirstein, J. M. Atienza, B. Xi, J. Zhu, N. Yu, X. Wang, X. Xu and Y. A. Abassi, *Assay and Drug Development Technologies*, 4, 545 (2006).

[110] M. Josephine, N. Y. Atienza, K. Shelli, B. Xi, X. Wang, X. Xu, A. Yama, and Y. A. Abassi, *Assay and Drug Development Technologies*, 4, 597 (2006).

[111] Y. *Abassi, Biochemica Roche Applied Science*, 2, 7 (2008).

[112] T. Nakagawa, Y. Nishi, C. Akihiro Kondo, Y. Yuko Shirai, C. Chie Honda, M. Asahi and T. Tanimoto , *Carbohydrate Research*, 346, 1792 (2011).

[113] J. Glamann and A. J. Hansen, *Assay and Drug Development Technologies*, 4, 555 (2006).

[114] S. B. Prakash and P. Abshire, Biosensors & Bioelectronics, 23, 1449 (2008).

[115] Y. Qiu, R. Liao and X. Zhang, *Analytical Chemistry*, 80, 990 (2008).

[116] Y. Qiu, R. Liao and X. Zhang, *Biophysical Journal,* 96, 1985 (2009).

[117] Y. Qiu, R. Liao and X. Zhang, *Biosensors & Bioelectronics*, 25, 147 (2009).

[118] H.Yang, Q. Wu, M. Tang, X. Liu, H. Deng, L. Kong, and Z. Lu, *Journal of Nanoscience and Nanotechnology*, 10, 561(2010).

[119] P. T.Witkowski, L. Schuenade, J. Wiethaus, D. R. Bourquain, A. Kurth and A. Nitsche, *Biochemical and Biophysical Research Communications*, 401, 37 (2010).

[120] A. W. Browne, J. L. Leddon, M. A. Currier, J. P. Williams, S. Jason , J. S. Frischer, M. H. Collins, H. Chong, C. H. Ahn, and T. P. Cripe, *PLoS ONE* 6, e19530 (2011).

[121] A. M. Vinggaard, E. C. B. Joergensen and J. C. Larsen, *Toxicology and Applied Pharmacology*, 155, 150 (1999).

[122] H. Yu and R. Jove , Nature Reviews: Cancer, 4, 97 (2004).
[123] A. K. Mankan and F. R. Greten, *Expert Opin Investig Drugs*, 20, 1263 (2011).
[124] H. Yu, M. Kortylewski, and D. Pardoll, *Nature Reviews: Immunol.* 7, 41 (2007).
[125] M. N. Chau, and P. P. Banerjee, *Biochemical and Biophysical Research Communications,* 377, 627 (2008).
[126] M. W. McCulloch, G. S. Coombs, N. Banerjee, T. S. Bugni, K. M. Cannon, M. K. Harper, C. A. Veltri, D. M. Virshup, and C. M. Ireland, *Bioorganic and Medicinal Chemistry*, 17, 2189 (2009).

In: Cell Proliferation
Editors: C. Zhang and X. Zeng

ISBN: 978-1-62417-352-3

Chapter VII

Cell Adhesion and Proliferation on Polymeric Biomaterials for Tissue Engineering

***Xiangqiong Zeng*[1,*] *and Changhong Zhang*[2]**

[1]Department of Surface Technology and Tribology,
Faculty of Engineering Technology, Institute for Biomedical Technology and Technical Medicine,
University of Twente, The Netherlands
[2]RandD Department, Molnlycke Health Care Company,
Wiscasset, Portland, Maine, US

Abstract

Synthetic polymeric biomaterials have been widely used in biomedical fields, such as prosthetic materials, dental materials, wound dressings, drug delivery devices, tissue regeneration implants etc. Polymeric biomaterials have exhibited high chemical versatility, and can mimic the natural properties of natural human tissues in biocompatibility, flexibility and physiological property. Moreover, biomolecular recognition of polymeric materials by cells has been obtained by surface and bulk modification via chemical or physical methods with immobilization of bioactive molecules.

This chapter discusses the synthesis, surface and bulk modification of the polymeric biomaterials for cell adhesion and proliferation control, and the further application in tissue engineering. Recent progress for the development of biomimetic materials to provide the biological cues for cell-matrix interaction to promote tissue growth in skin, nerve, cardiovascular and bone tissue engineering are also described.

[*] Email address: x.zeng@utwente.nl.

1. Introduction

Tissue engineering, an important emerging topic in biomedical engineering, has shown tremendous promise in creating biological alternatives to replace biological tissues and organs for a wide range of medical conditions involving tissue loss or dysfunction, for instance, skin, nerve (retina, peripheral nerves, spinal cord), cardiovascular (cardiac muscle, blood vessels, heart valves), musculoskeletal (bone, cartilage, muscle, ligament, tendon, intervertebral disc), digestive and endocrine (liver, pancreatic islets, intestines, oesophagus, parathyroid glands) [1-2]. Therefore, the field of tissue engineering places complex demands on the materials it uses. The materials chosen to support the intricate processes of tissue development and maintenance need to have properties which serve both the bulk mechanical and structural requirements of the target tissue, as well as enabling interactions with cells at the molecular scale.

1.1. The history of Biomaterial Development

Biomaterials have been used for a variety of biomedical applications for thousands of years. As described by P. Roach [3], one of the first known attempts to repair damaged tissues was carried out by the ancient Egyptians approximately 4000 years ago, who used plant and animal derived fibers to sew wounds. Wooden prostheses have been found attached to the mummified remains of a human, attempting to rebuild a foot following the amputation of a toe [4]. Metals came into use in dental practice as early as the first or second century AD with wrought iron replacement teeth being cast from the original and impacted into the jaw [5]. Although these now appear to be crude attempts to replace damaged body parts, the choice and development of materials used for implantation has only relatively recently become a science, and biomaterials have been of growing importance in various biomedical technologies.In general, the development of biomaterials can be summarized into three generations as shown in table 1.

Table 1. The development of biomaterials [3]

Generation	Characteristics	Examples
1	Developed in the 1960s; Prevent any adverse response of the host to the implanted material, i.e. bioinert; Materials in contact with blood necessitate the use of anti-coagulants.	Polytetrafluoroethylene (PTFE) venous implants [6].
2	Resolve implant-tissue interface problems like foreign body response and ultimate rejection of the implanted material; Widely used by the mid 1980s.	Bioglass^{TM}, promote the formation of mineralized bone following attachment of osteoblasts [7]; Bioresorbable materials, e.g. copolymer of glycolic acid (PGA) and L-lactic acid (PLLA), allow tissue in-growth and regeneration to replace implanted biomaterial by host generated tissue [8].

Generation	Characteristics	Examples
3	The term "tissue engineering" generated by Langer and Vacanti in 1993 [9]; Stimulate specific cell responses, incorporating cell and activating components to encourage tissue regeneration; Incorporate cells into scaffolds ex-vivo mimicking natural tissue, then implanted into a host to repair or replace damaged tissue; Injectable scaffolds allowing non-invasive delivery of support material, cells and other additives like drugs; Bioresorbable.	Polymeric materials functionalized with sugars, peptides and other moieties to generate structures capable of specific binding interactions and the development of intelligent delivery vehicles [10-11]; Tissue engineering scaffolds: hydrogel, fibrous, custom, porous [1]; Injectable poly(lactic-co-glycolic) acid (PLGA) scaffolds [12], in situ-forming hydrogels [13], injectable polymer, poly(propylene fumarate) (PPF), as bone cement [14].

1.2. Parameters for Biomaterial Research

For the construction of advanced bioartificial tissues and organs, biomaterials should not just be passively tolerated by the cells, but should actively promote specific cell responses in a controllable manner. Generating optimal biomaterials, that can be used to control cell-microenvironment interactions, and to regulate the extent and the strength of cell adhesion, the subsequent cell proliferation and differentiation, is very challenging.

The manner in which biomaterials interact with cells depends strongly on the physical and chemical properties of the biomaterials, and the microenvironment [15-16].

Firstly, properties of materials can be regulated based on their physical and chemical properties. Physical characteristics generally refer to basic properties, like stiffness /pressure/elasticity, viscosity, porosity, crystallinity, and mechanical properties including swelling, permeability, stress strain properties and density, which affect the mechanical and macroscopic features of the materials. Chemical properties can modify the biological responsiveness, which can be modulated by derivatization such as bioactive substitution, modification, hydrophilicity and chemical responsiveness, like electrostatic responses and external stimuli (i.e. pH, temperature, light sensitivity, humidity) [17].

Secondly, in addition to the biomaterial optimization, various environmental factors are also very important for the design of biomaterials. Because cellular behavior is regulated by extrinsic signals from the microenvironment such as cell-cell contact, cell-extracellular matrix (ECM) interactions, and cell-soluble factor interactions, the integration of biomaterials and cell microenvironment can be affected by cell type, medium supplements, ECM and spatial organization (2D/3D) [16, 18].

The microenvironment of cell adhesion, proliferation as well as differentiation can be modified by the addition of bioactive components such as ECM molecules, environment conditions (i.e. temperature, pH), and soluble signaling molecules [19].

The complexity of the interacting parameters makes design of biomaterials a very exciting while challenging research. However, it is possible to select certain critical influences according to specific application and present them to cells in order to attempt to control behavior. Tremendous amount of valuable work has been done in this field. It is significance to make a comprehensive summarization of those achievements.

1.3. The Advantage of Synthetic Polymeric Biomaterials

The first issue with regard to tissue engineering is the choice of suitable material. As introduced in table 1, the desirable characteristics of these materials are biocompatibility, biodegradability and desirable cell responses. Over the past century, many biomaterials with these characteristics, including natural polymers, synthetic polymers, ceramics, metals, and combinations of these materials, have been investigated, as exhibited in table 2. Among which, metals and ceramics have two major disadvantages for tissue engineering application, e.g. limited processability and non-biodegradable, except for biodegradable bioceramics such as α-tricalcium phosphate, β-tricalcium phosphate. For these reasons, polymeric materials have received an increasing amount of attention from scientific and medical communities [2].

Table 2. Materials for tissue engineering

Category	Some examples	Some applications	Reference (Ref.)
Metals	Stainless steels, cobalt-based alloys, titanium-based alloys	Orthopedic tissue replacements, hip endoprosthesis, bone regeneration	[2, 20]
Ceramics	Alumina, zirconia, calcium phosphate, bioglass		
Natural polymers	Collagen, glycosaminoglycan, starch, chitin, chitosan, silk fibroin, fibrinogen, gelatin, cellulose derivatives, xyloglucan, agarose/alginate, kerateines	Repair nerve, skin, cartilage, bone, sealant for vascular prostheses, carriers for drug delivery, wound dressings	[2, 13, 20-26]
Category	Some examples	Some applications	Reference (Ref.)
Synthetic polymers	Biocompatible: Poly(ethylene glycol) (PEG), poly(vinyl alcohol) (PVA), poly(acrylic acid) (PAA), poly(2-hydroxyethyl mechacrylate) (PHEMA), poly(methyl methacrylate) (PMMA); Biodegradable, approved by FDA: poly(α-hydroxy acids), e.g. poly(lactic acid) (PLA), poly(glycolic acid) (PGA), poly(caprolactone) (PCL) and their copolymers; Alternatives: polycarbonate, PPF; poly(ester amide), poly(urethane) (PU); poly(anhydride); poly(orthoester); poly(phosphazene); poly(glycerol sebacate); poly(phosphoester)	Bone cements, surgical sutures, stents, wound dressings, hard and soft tissue engineering applications (skin, bone, cartilage, tendon, ligament, nerve, lung, heart, blood vessels, skeletal muscle), drug delivery	[1-2, 20, 23-24, 27-28]
Combinations	gelatin/PCL composite; PLGA reinforced with hydroxyapatite (HA) short fibers	Filler material for bone defect repair and as artificial bone matrix	[2, 21, 29]

Natural polymers may most closely simulate the native cellular milieu, while the main limitations for their wide applications are large batch-to-batch variations upon isolation from biological tissues, poor mechanical performances and also poor processability [2, 21].

Synthetic polymeric biomaterials have been widely used in biomedical fields, such as prosthetic materials, dental materials, wound dressings, drug delivery devices, tissue regeneration implants etc., although there are also some limitations of these materials, for instance, all polyesters release acidic degradation products that can adversely affect biocompatibility [2], and some of these polymers are hydrophobic that can cause poor wetting and lack of cellular attachment and interaction. They typically have exhibited high chemical versatility so that the mechanical and chemical properties can be tailored; they can be non-toxic, reliable, readily available and relatively inexpensive to produce, and in many cases can be processed under mild conditions that are compatible with cells; they can mimic the natural properties of natural human tissues including biocompatibility, flexibility and physiological property; moreover, biomolecular recognition of polymeric materials by cells can be obtained by surface and bulk modification via chemical or physical methods with immobilization of bioactive molecules [1, 22].

In this chapter, we discuss the synthesis, surface and bulk modification of the polymeric biomaterials for cell adhesion and proliferation control, and the further application in tissue engineering. Recent progress for the development of biomimetic materials to provide the biological cues for cell-matrix interaction to promote tissue growth in skin, nerve, cardiovascular and bone tissue engineering are also described.

2. The Synthesis of Polymeric Biomaterials and the Effects on Cell Adhesion and Proliferation

Incorporating multiple physical, chemical and biological functionalities to biomaterials will yield synthetic polymeric biomaterials which mimic the natural cell environment more closely and participate in a dynamic, bidirectional exchange of information with cells. The synthesis of polymeric biomaterials by varying monomer ratios, incorporating various functional chemical groups and bioactive molecules will result in different physical and chemical properties, which can modulate the surface and bulk properties of biomaterials and in turn regulates the adhesion and proliferation of a variety of cell types.

2.1. The Design of Polymer Molecular Structure

The varying of monomer ratios in copolymers usually results in the changes of hydrophilicity, and behaviors of cultured cells on biomaterials might differ to a great extent depending on the hydrophilicity of the biomaterials. Pei-Jen Lou, *et al.* [30] identified distinct cell adhesion and proliferation of human diploid fibroblasts (HDFs) on poly(ethylene-co-vinyl alcohol) (PEVAL) consisting of various ratio of hydrophilic vinyl alcohol segments and hydrophobic ethylene segment. PEVAL44 (containing 44 mole% ethylene) and PEVAL27 (containing 27 mole% ethylene) membranes were prepared and HDFs were cultured on the membranes. The work shows that the decreasing of hydrophobic property of PEVAL induced characteristic senescence-associated phenotypic changes such as larger cell shape, re-organized actin cytoskeleton, lower proliferation capacity, higher levels of senescence-associated β-galatosidase (SA β-gal) activity, and upregulation of the cell-cycle inhibitor p53 and its transcriptional target p21 in the cultured HDFs. The author of this chapter, Changhong Zhang has done comprehensive research on biodegradable polyurethanes (PUs) [31]. A series of degradable PU-based light-curable elastic hydrogels were synthesized from polycaprolactonediol (PCL-diol), PEG, lysine diisocyanate (LDI), and 2-hydroxyethyl methacrylate (HEMA) through UV light initiated polymerization reaction. By changing the PCL to PEG ratio during the prepolymer synthesis, PUs with different soft segmental structures, hydropilicity and cytophilicitywere obtained after light-initiated polymerization. Mouse embryonal carcinoma-derived clonal chondrocytes were used to study the cytocompatibility of the synthesized polymer. It was found that chondrocyte attachment, proliferation rates varied with changes in the PCL/PEG ratio. With a higher PEG ratio, lower cell attachment and proliferation were observed [32]. While in his another study, two types of PUs were synthesized from methylene di-*p*-pheynyl-diisocyanate (MDI), PCL-diol, and chain extenders of either butanediol (BD) or 2,2'-(methylimino)diethanol (MIDE). PU containing MIDE is more hydrophilic and retains more liquid during in vitro culture. Preliminary cytocompatibility studies showed that fibroblasts adhere better and proliferate faster on MIDE containing PU than BD containing PU [23]. Therefore, it is difficult to draw a conclusion on the effect of hydrophilicity. More detailed discussion will be made in session 3 (surface modification).

In addition to hydrophilicity, mechanical properties of polymer substrate could also play a critical role in determining cell adhesion and proliferation. Polymers can have tunable elastic stiffness by varying the ratio monomer to cross-linking agent. For instance, polyacrylamine (PA) gels can be prepared having elasticity in the range 0.1-100kPa, which can be adjusted to mimic hard tissues such as bone (E~30kPa) through to soft tissues such as brain (E~0.5kPa); other materials were also used to extending this range such as polydimethylsiloxane (PDMS) (E~10-1000kPa) and collagen gels (E~0.001-1kPa) [3, 5, 33-34]. A systematic investigation has been performed by Cai, *et al.* on regulating materials properties and cell behavior using hybrid networks composed of amorphous PPF and three poly(ε–caprolactone) diacrylates (PCLDAs) with variance in crystallinity and melting temperature [35]. With the increasing of PCLDA composition, the crosslinked PPF/PCLDA changes from amorphous to semi-crystalline. Cell studies were conducted on crosslinked polymer disks, using both mouse MC3T3-E1 cells and rat Schwann cell precursor line (SpL201) cells for bone and nerve tissue engineering applications. It was observed that the role of cystallinity in strengthening materials consequently enhanced cell attachment and proliferation. While in the studies on another semi-crystalline polymer, PLLA, cell responded differently to semi-crystalline and amorphous samples, and the nanometer-scale roughness was proposed to interpret the enhancement of osteoblast proliferation [36-37]. A substrate having a gradient change in rigidity was fabricated by adding two droplets containing varying ratios of acrylamide and crosslinker onto a glass substrate adjacent to each other [38]. The chemistry of the surfaces was kept constant apart from the varying crosslinkers that may have been presented at the surface, allowing cell-surface interactions to be assessed in terms of the stiffness parameter. The migration of 3T3 fibroblasts was observed, with cells found to move onto the more rigid portion of the substrate if they were initially adhered to the soft region near the border. Extensive studies have been performed on the role of material stiffness in determining cell spreading, proliferation and differentiation for nerve tissue engineering application [39-45]. And it was found that soft substrates can stimulate neurite extension and branching, while inhibit glial cell spreading and proliferation; a range of substrate stiffness (E: 0.01-10kPa) was established to which neural stem cells respond. Since different cells respond different to the same substrates, it is still unclear if there exists a critical value for surface stiffness to be sufficiently high for one certain cell type to attach and proliferate on a certain substrate.

The charge characteristic of polymeric biomaterials is another factor influencing cell behavior. A sulfonic acid containing PU (MP530B) was synthesized from MDI, PCL-diol and N,N-bis(2-hydroxyethylhydroxyethyl)-2-aminoethane-sulfonic acid (BES) as the chain extender. A non-sulfonic acid incorporating PU (MP530M) , which had a MIDE chain extender was used as a comparison because it has a similar chemical structure to MP530B but contains no sulfonic acid group as shown in figure 1 [46].

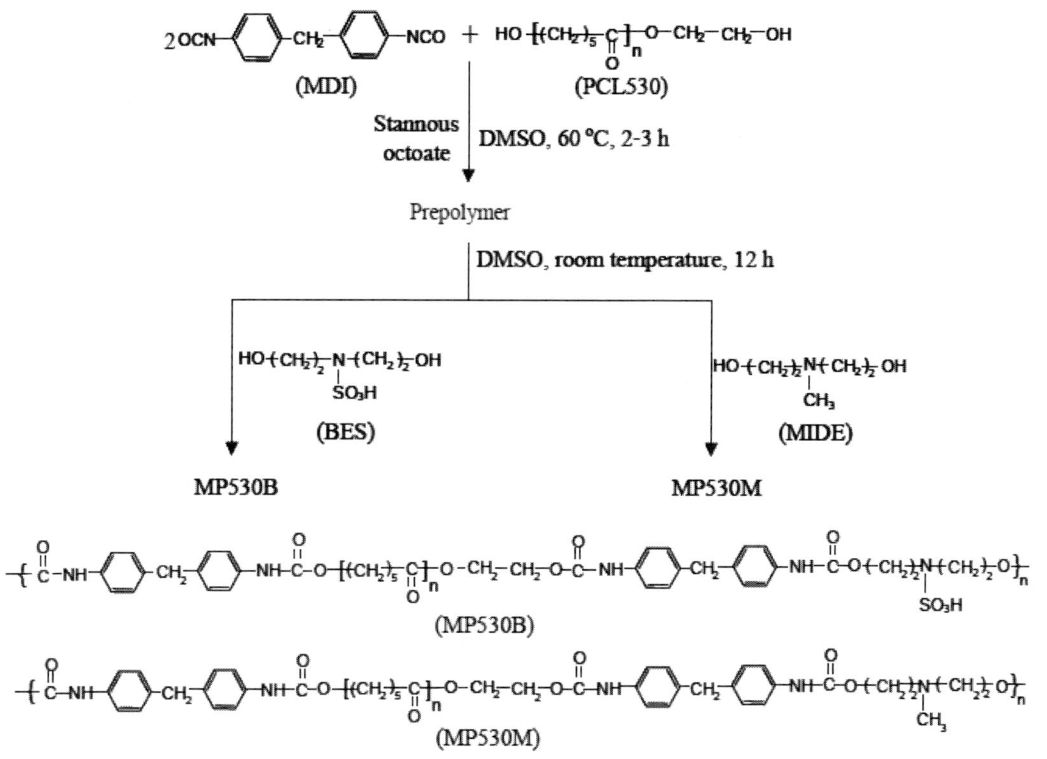

Figure 1. Synthesis scheme for the preparation of MDI, PCL530, MIDE or BES based Pus.

Fibroblast cells were cultured on the polymer surfaces. Results indicated that MP530B exhibited much lower cell attachment and proliferation than MP530M. The low cell proliferation rate of fibroblasts on MP530B is most likely attributed to the existence of negative sulfonic acid groups.

In cell culture, the sulfonic acid on the materials surface can be ionized by culture media (pH=7.4), and because the flexibility of the polymer chains, the ionized sulfonic groups can form a thin negatively charged layer on the substrate surface. This layer may repel the fibroblast attachment by the electrostatic interaction. In MP530M, the tertiary nitrogen atoms in chain extender MIDE can form cationic groups in the water and five a slightly positive polarity on the polymer chains.

In culture media the polymer chains can change the conformation to form a positively charged layer on the substrate, which will interact with negatively charged cell membrane surface and most serum proteins, thus promote cell attachment and proliferation.

2.2. The Incorporating of Bioactive Molecules

The cell-biomaterial interaction in the aforementioned approaches is mainly non-specific cell-material interactions via weak chemical bonding, such as hydrogen bonding, electrostatic, polar or ionic interactions between various molecules on cell membrane and functional chemical groups on the polymers.

To actively promote specific cell responses with biomaterials in a controllable manner, another approach is to incorporate biosignals to create a controlled, bioinspired extracellular environment to direct tissue-specific cell response. When presented with appropriate biological cues, cell receptors will bind to these signaling biomolecules and transmit the signals intracellularly by activating signaling cascades.

These cascades will modulate gene expression and determine important cell fate processes such as proliferation and differentiation to ultimately regenerate functioning tissue, which is a powerful tool for enhancing cell-biomaterial communication and inducing desired cell behavior [1, 47-48].

The molecular mechanism of receptor-mediated binding between cell and biomaterials is displayed in figure 2. Generally, it includes three steps, the adsorption or incorporating of bioactive molecules to biomaterials, the ligand binding between the adhesion motif of bioactive molecule and the integrin receptor of the cell, the focal adhesion between integrin and structural/signaling molecules in the cell.

Some studies have incorporated bioactive molecules or short adhesion motifs derived from binding regions of bioactive molecules into polymer backbone during synthesis. One example is hydrogels crosslinked by protease cleavable peptides and decorated with biological functionalities including adhesion ligands.

These can be modular systems in which any combination of bioactive molecules can be added to a bioinert polymer background [1]. The synthetic process used by the Hubbell group allows any thiolated proteins or peptides to be included in PEG hydrogels by reaction with bis(α,ω-vinylsulfone)poly(ethylene glycol) before crosslinking [49].

The author of this chapter, Changhong Zhang incorporated a short peptide, Arg-Gly-Asp-Ser (AGSR) into a PU-based hydrogel during the light-curing process. AlamarBlue assays on the chondrogenic cell attachment and proliferation showed PU with immobilized AGSR in the hydrogel network exhibited significantly higher cell attachment and proliferation than that of PU-based hydrogel.

There are more investigations on surface and bulk modification with bioactive molecules, which will be introduced in session 3 and 4.

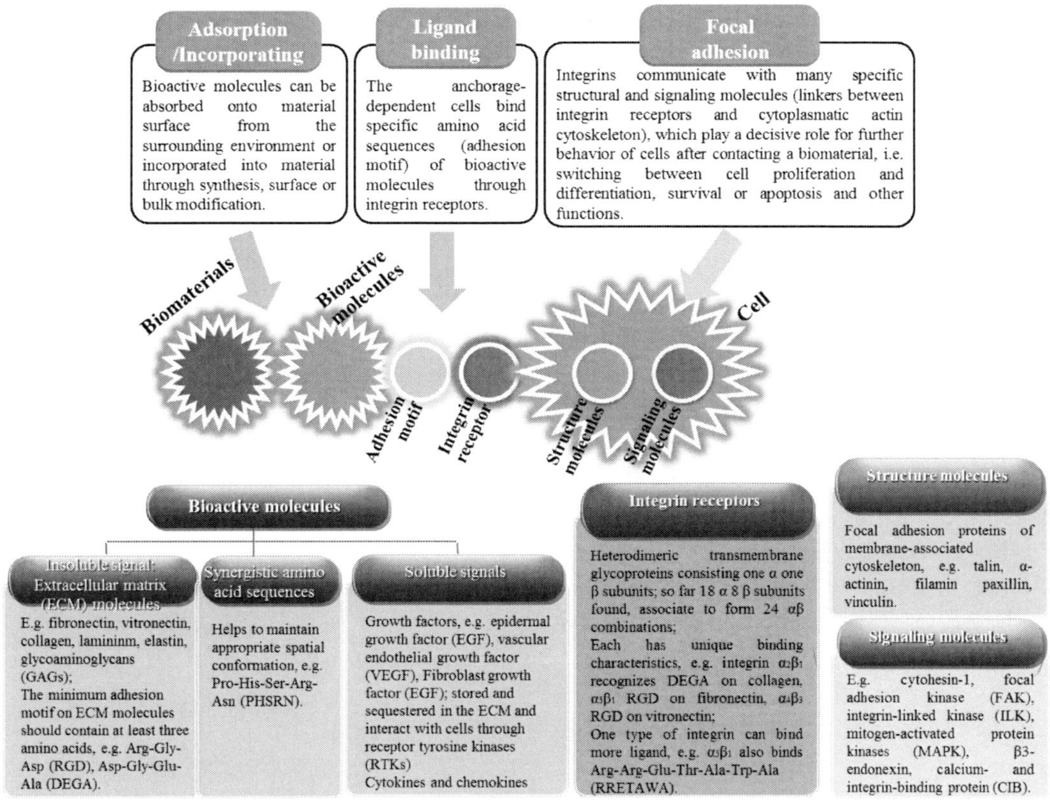

Figure 2. Molecular mechanism of cell-biomaterial interaction.

3. The Surface Modification of Polymeric Biomaterials and the Effect on Cell Adhesion and Proliferation

Cell-material interactions are essential to the tissue-engineering applications of biomaterials. Despite many synthetic biomaterials having physical properties that are comparable or even superior to those of natural body tissues, they frequently fail due to the adverse physiological reactions they cause within the human body [50]. Moreover, the response of cells to biomaterials is often initiated by cell contact and adhesion to the biomaterial surface [16].

Surface modification of biomaterials is an economical and effective method by which biocompatibility and biofunctionality can be achieved while preserving the favorable bulk characteristics of the biomaterial, like strength and inertness [50]. Numerous surface modification techniques have been developed that enable tailoring of the surface morphology, structure, composition, and properties of the material to a specific need, including plasma treatment [50-51], lithography, electrospining, dip-coating, spin coating, layer-by-layer electrostatic assembly [3], chemical etching and casting [1, 52], vapor based polymer coating, ion sputtering [53], micro- and nano-patterning [53-56], photochemistry [57], etc.

Correspondingly, clear characterization of the chemical compositions and physical structures of the biomaterial surface has profound scientific importance, leading to insight understanding of cell-biomaterial interactions [53]. A wide and varied surface analytical approaches has been applied to study the interaction between biology and man-made materials, including contact angle measurement, X-ray photoelectron spectroscopy, mass spectrometry, atomic force microscopy, surface plasmon resonance, electron microscopy, visible adsorption, fluorescence microscopy [3], scanning electron microscopy [28], ellipsometry, attenuated total reflectance Fourier transform spectroscopy, quartz crystal microbalance [53], etc.

Based on these surface modification and characterization techniques, pioneer scientists have achieved great understanding on the roles of the surface physicochemical characteristics of biomaterials in determining cell behavior in the past several decades [35]. The key material characteristics identified to be important for biological recognition are surface topography, chemistry and mechanical properties, which may also correlate together to influence cell behavior [3, 35, 58].

In this session, the effect of surface topography and surface chemistry of synthetic polymeric biomaterials on cell adhesion and proliferation will be discussed for the application in skin, nerve, vascular and bone tissue engineering.

3.1. Surface Topographical Strategy and Its Application

Cells inhabit a complex environment rich in topographical and physicochemical cues at length scales from several nanometers to microns [1]. It has been shown that surface topography of biomaterials influences cellular behavior including adhesion, morphology (orientation and polarization), proliferation and differentiation [3]. Therefore, topographic modulation of tissue response can be one of the most important considerations during the design and manufacture of biomaterials [55].

3.1.1. Surface Topographical Strategy

The structural organization of tissues plays a major part in deciding the degree and direction of tissue growth and cell movement. Various studies have indicated that it may be possible to design the surface texture of implanted materials to improve the performance of an implant [55].

The main parameter of surface texture influencing cellular response is surface roughness. The most widely used parameter for characterizing surface roughness is Ra(the average peak-to-valley height). However, this measure does not give any record of the type of surface topography, because not only the height and depth of the surface irregularities are important, the spacing between the irregularities, their different shapes, e.g. pyramids, cones, ridges, grooves, round pores, the curvature of the valleys, and the sharpness of the peaks on the material surface are also important and could hamper cell adhesion and spreading [47, 54].

In addition, the cellular response to the surface topographies is also influenced by the dimensions of the surface features, ranging from nano, sub-micron to micro-scales, depending on cell types, cell-cell interaction, as well as substrate composition and topography type. Therefore, the parameters of the surface roughness have their optimum range, which differ for each type of material (i.e. carbon-based composites, metals, ceramics, synthetic polymers), and for each type of cells [53].

Advances in mciro-fabrication technologies allowed greater control over surface morphology[59]. A widely used technique is engineering micropatterned surfaces by varying the material surface roughness and topography, i.e. creating hollows and prominences of various sizes, shapes, spacing and distribution, such as grooves and ridges, pits, pillars, boxes, cylinders or honeycombs [54].

Microgroove/ridge surfaces have shown significant control over cellular behaviors. The most important phenomenon is that the cell spreading, alignment, and migration can be oriented along the grooves/ridges [53]. One theory that accounts for this is called "contact guide effect", in which cell integrin receptors in focal contact transfer the variable degrees of tension or compression into the cytoskeleton, and cell stretch receptors subject to these stresses will be activated and reorganize the cytoskeleton according to the surface topography [60-62]. Another explanation is the changes of surface free energy due to the edges and disruptions may be the reason for the cell orientation [63]. It was found that rectangular ridges with 5μm width (1-5μm high) enhanced the adhesion of a kind of marine spore, which decreased the adhesion of endothelial cells [64].

Surface microroughness is a controversial factor affecting the behavior of cells on artificial materials. The cells typically studied on these materials, i.e. anchorage-dependent mammalian cells of various tissues and organs, including vascular tissue or bone, are usually between 10μm and 50μm in diameter, if they are in suspension, where they acquire a rounded shape. When adhered and spread on the material surface, their spreading area can reach from several hundreds of to several thousands of μm^2. Thus, these cells are inherently sensitive to the microtopography of their environment, and many studies have reported that microroughness significantly affected the cell response to the material. Some studies have reported a positive influence of surface microroughness on cell adhesion, growth and maturation, whereas in other studies this influence has been considered negative [54]. The mechanism of this dual effect of surface microroughness on cell colonization still remains unclear, which may be related to the difficulty in comparing the data from different research group due to the surface roughness, the changing of other surface physical and chemical properties, e.g. energy and wettability when changing the surface topography [54], and the different cell types employed in the studies [65].

In order to compare between topographic studies, an attempt to standardize the investigation of grooved structures by aspect ratio was tried by Crouch *et al.* [3, 66]. An increase in fibroblast cell alignment and polarization was found with increasing aspect ratio, with 80% alignment being induced down to a ratio of ~0.05 (depth 0.1μm, width 2.0μm). Further comparison of the data to that of others by readjusting the aspect ratio, they found similar trends despite varying cell types and feature types. Others have likewise found that aspect ratio of the surfaces was important for cell orientation, with fibroblasts aligning well to 1μm grooves/ridge structures but not to 2, 10 and 20μm patterns, all being 0.5μm deep [67]. It is however suggested that once maximum alignment is achieved further increase in feature depth has no effect [68].

The physicochemical properties of the material surface can be varied not only on a microscale level but also on a nanoscale level. The principles for creating nanoscale surfaces are similar to those used for surface micro-fabrication [54, 69].

Nanoscale roughness of the material surface has been unambiguously considered as a desirable factor that has a positive influence on the adhesion, growth and maturation of cells. The reason is that the nanostructure of a material resembles the nanoarchitecture of the natural ECM, e.g. its organization into nanofibers, nanocrystals, nano-sized folds of ECM molecules.

On nanostructure surfaces, the cell adhesion-mediating ECM molecules therefore adsorb in an appropriate geometrical orientation which gives cell adhesion receptors access to specific sites in ECM molecules, such as amino acid sequences like RGD, which serve as ligands for these receptors [54, 70]. Upon a backdrop of fibrous ECM proteins with nanoscale diameters, the specific binding domains of proteins and growth factors present numerous sites for interaction with cell surface receptors, while the cleavage sequences of proteases await scission by cell-derived enzymes. In this way, cells interact entirely with a landscape of nanoscale physical features. As cells move over a substrate they extend and retract filopodia peppered with integrin receptors, effectively 'feeling' the environment as they migrate.

Many cell types respond directly to nanoscale surface features by changing their phenotype or activity and it has been seen that a number of cell types preferentially align along grooved surfaces [8-93, 8-94]. In general, the cell behavior on nanopatterned surfaces can be regulated by the orientation, shape and size of the nanoscaleirregulatities on the material surface.

Firstly, the orientation of the nanofibers in nanofibrous scaffolds, i.e. random or aligned, modulated the adhesion, spreading, shape and cytoskeletal organization of various cell types. The cell proliferation, differentiation, phenotypic maturation and functioning were usually better on aligned nanofibers than on randomly oriented nanofibers or on unpatterned flat surface. For instance, composite nanofibers of multiwalled carbon nanotubs and PU aligned in parallel stimulated proliferation and collagen secretion in human umbilical vein endothelial cells (HUVECs) [73]; in cultures of rat bone marrow stromal cells on PLLA nanofibers, the calcium content on fibers aligned in parallel was significantly higher than that of randomly oriented fibers [74]; in neural stem cells derived from rat hippocampus and cultured on aligned PCL fibers, a higher fraction of cells exhibited markers of neuronal differentiation after stimulation with retinoic acid than cells on randomly distributed fiber [75]. The better performance of cells on aligned nanofibers is a typical example of morphological and then biochemical control of the cell behavior by the adhesion substrate. Aligned nanofibers guide the cells to be elongated along the major fiber axis. The elongated cell shape then stimulates specific biochemical pathways of extracellular signal transduction. Secondly, the nanofibrous shape of nanoscale irregularities on the material surface is considered to be more advantageous for cell performance than other types of irregularities in general. An explanation is that nanofibrous structures bear a better resemblance than other types of nanoscaleirregulatities to the natural ECM, which is often fibrous in character, e.g. collagen [54]. In addition, the surface curvature is also important [3]. For example, albumin retains a more native-like structure on surfaces presenting high surface curvature (having a radius of curvature ~7nm) fibrinogen is heavily structurally distorted, possibly due to the rod-like protein becoming wrapped around the topography on adsorption. In contrast, the opposite trends are observed for both proteins on surfaces presenting lower surface curvature (up to radius ~83nm), with albumin becoming more denatured with decreasing surface curvature and fibrinogen becoming more native-like. This is because albumin is a globular protein measuring ~14x4nm whilst fibrinogen is rodlike measuring ~40nm along its long axis and having a diameter of ~4 nm [76]. Thirdly, the size of the nanoscale irregularities on the material surface can also manipulate cell behavior. For example, the attachment and spreading of human fetal osteoblastic cells (hFOB) were greater on polymeric surfaces with shallower pits (14 and 29nm) than on surfaces with deeper pits (45nm) [77]; while on polymeric surfaces with nanoislands 11, 38 and 85nm in height, the proliferation of hFOB cells and also their osteogenic differentiation correlated inversely with the height of the nanoisland [78]; moreover, nanostructured surfaces promote preferential adhesion of osteoblasts over other cell types, such as fibroblasts, chondrocytes and smooth muscle cell. This has been explained by the preferential adsorption of vitronectin on nanostructured surfaces, due to its relatively small and linear molecule (15nm in length) in comparison with other larger and more complicated ECM proteins, e.g. laminin, the configuration of which is cruciform and 70nm both in length and in width [3, 79-80].

3.1.2. Application in Nerve Tissue Engineering

Damage to the nervous system through crushed or transected nerve tracts, caused by mechanical, thermal, chemical, or ischemic factors, can impair various nervous system functions by the interruption of communications between nerve cell bodies and their targets. Without appropriate interventions, injury sites are generally occupied by dense scar tissue. Although axons are free to be rerouted around the scar region, this may be at the expense of losing connections with their intended targets. Therefore, helping regenerating axons cross the lesion area and guiding them to appropriate targets represents an important step in the early regenerative process, in which bioengineering approaches may be effective [81-82].

One way to help is by creating an artificial growth permissive substrate to connect the gap between damaged nerve tracts across the scar. Such a substrate, with the two ends spanning across the lesion gap, is defined as a bridge or bridging substrate. Some bridges have been shaped into neural guidance channels with tubular structures, which have been developed from several material types with the aim being to direct axonal growth along an implanted scaffold in order to bridge the damaged region at a nerve injury site. Recent studies further advance earlier works by bridging progressively larger gaps of both the central and peripheral nervous system as shown in table 3 [3].

Table 3. Surface topography and the application in nerve tissue engineering

Polymeric material	Surface topography	Cell adhesion and proliferation	Ref.
Aromatic polyether based PU or PLGA hollow fiber membranes (HFMs)	Highly aligned textures on the inner surface of semipermeable HFMs.	Both alignment and outgrowth of regenerating axons increased significantly on HFMs with aligned textures compared to those with smooth inner surface.	[81]
Poly-D-lysine (PDL) coated PCL filament; PDL/laminin coated PCL filament	Longitudinal microgrooves; circumference 1.5 times as large to concentric filaments (104.5μm vs. 69μm).	Displayed good cell attachment, supported Schwann cell proliferation as well as guided axonal outgrowth. Schwann cells were confined to lateral surface areas instead of growing in the depths of constricted grooves.	[82]

3.1.3. Application in Vascular Tissue Engineering

The largest cause of mortality in the Western world is atherosclerotic vascular disease. This comes in the form of coronary artery and peripheral vascular disease.

Current treatment methods involve bypassing occluded arteries with either autologous veins or biocompatible synthetic materials [52]. While Dacron and PTFE have had success in replacing large arteries, smaller diameter arteries, namely those under 6mm, have seen very limited success [83]. Thus, new synthetic material formulations for small diameter vascular grafts are needed.

Currently, most polymers used by medical manufacturers to construct vascular grafts consist of topographies that are much different than those found in vascular tissue [84]. For instance, it has been shown that tubular scaffolds of PGA can support vascular smooth muscle and endothelial cell growth in vitro [85-86].

However, these materials are composed of micron sized fibers and micron surface topographies that are dissimilar to the natural vascular tissue because the fundamental structural components of the blood vessel wall are on the order of nanometers [87-88].

Table 4. Surface topography and the application in vascular tissue engineering

Polymeric material	Surface topography	Cell adhesion and proliferation	Ref.
Poly(epsilon-caprolactone) films	Micropatterned with honeycomb-shaped pores.	Procine aortic endothelial cells: three times higher expression of focal adhesion kinase autophosphorylated at the tyrosine residue (pFAK) than on a corresponding unpatterned film (fibronectin adsorbed preferentially on the pore edges).	[89]
PU	Microstructures with arrays of parallel channels 95μm in width and 32μm in depth.	The retention of endothelial cells was markedly improved from 58% on non-patterned surfaces to 92%.	[90]
PLLA	Patterned with grooves and submicron- and micron-scale ridges from 350nm to 1750nm in width.	Endothelial cells (line HUVEC and CPAE): the highest cell adhesion strength was found on surface of ridges 700nm in width and grooves 350nm in width.	[91]
Polyamidnanofibers	Nanofibers aligned in parallel or randomly.	Constructing bioartificial heart valves: valve interstitial cells (VIC) cultured on randomly oriented nanofibers were flat and polygonal, while on nanofibers aligned in parallel were spindle-shaped, oriented along the fibers; their α-actin-containing filaments arranged randomly forming a mesh-like structure on the randomly aligned nanofibers, and arranged in parallel with the long axis of the cells on the parallel aligned nanofibers.	[92]
PLGA	Nanostructured.	Engineering bioartificial vascular grafts: the numbers of initially adhering rat vascular endothelial and smooth muscle cells, their proliferation activity and final population densities were higher on nanostructured PLGA thanon flat PLGA.	[93]
PLGA	Nanostructured topography by using polymer/elastomer casting methods	Endothelial and smooth muscle cell densities increased on nanostructured cast PLGA.	[52]

Therefore, it is hypothesized that the cellular response important for vascular tissue regeneration will be enhanced on synthetic polymeric biomaterials through the creation of micro- and/or nano-scale surface topography mimicking the natural surface architecture of the vascular wall. The progress of this approach is described in table 4.

3.1.4. Application in Skin and Bone Tissue Engineering

Surface topography varies widely among commercially available orthopedic and dental implants. While it is generally accepted that the surface topography of an implant influences the formation of skin and bone, and affects its performance, a few studies have dealt with this important feature as demonstrated in table 5.

Table 5. Surface topography and the application in skin and bone tissue engineering

Polymeric material	Surface topography	Cell adhesion and proliferation	Ref.
Fibronectin-preadsorbed Silicone	Textured with 2, 5, 10μm wide grooves (2, 5, 10MU, respectively).	Human fibroblasts: Cells on smooth surface went faster into the S phase than cells on textured silicone. Cells on 10MU proliferated less than cells on 2 and 5MU.	[94]
Titanium-coated epoxy	Different micromachined grooved or pitted surfaces ranged between 30 and 120μm.	Fibroblasts oriented along 3 and 10μm deep grooves whereas they inserted obliquely into 22μm deep grooves. Bonelike foci were oriented along the long axis of the grooves.	[95]
Polystyrene (PS) implantable disk	Either smooth or microgrooved (1–10μm) on both sides.	Subcutaneously in a goat: 1μm deep and 1-10μm wide microgrooves did not influence tissue response around these PS implants in soft tissue.	[96]
Silicone	Parallel surface grooves with a groove and ridge width of 2, 5, and 10μm. The groove depth was approximately 0.5μm.	Microfilaments and vinculin-containing attachment complexes of rat dermal fibroblasts (RDF): more oriented along 2μm grooved substrate than on 5 and 10μm grooved surfaces. Vinculin was located mainly on the surface ridges on all textured surfaces. In contrast, bovine and endogenous fibronectin and vitronectin were oriented along the surface grooves.	[97]
Chitosan/poly-L-lysine films	Fiber-like, particle- or granule-like irregularities.	The spreading, assembly of actin cytoskeleton, proliferation, expression of osteocalcin and mineralization of MC3T3-E1 cells were more pronounced on surface with fiber-like morphologies than with particle-like irregulatities.	[98]
PLA or PS	Microgrooves with depths of 0.5, 1.0, 1.5μm and widths of 1, 2, 5 and 10μm	Enhance mineralized ECM production and cell alkaline phosphatase activity of rat bone marrow cells than on smooth surface.	[99]
PDMS	Micro-pillars with 6μm in height and 5, 10, 20, 40μm in diameter	Promote human bone marrow-derived connective tissue progenitor cells' spreading and adhesion.	[100]

3.2. Surface Chemical Strategy and Its Application

Biomaterial surface chemistry has been shown to influence cell adhesion, cytoskeleton organization, and different signaling cascades to regulate cellular morphology, migration, proliferation, differentiation, survival and function. These studies provide invaluable fundamental information to the further development of regenerative medicine and tissue engineering [30].

When foreign materials come into contact with body fluid or cell culture medium, the initial response is protein adsorption onto the materials' surfaces within a small fraction of a second. Thus, the materials interact with the cells through the adsorbed protein layer [3, 101]. The identity, concentration and conformation of proteins adsorbed to the surface from the complex serum mixture play critical roles in determining subsequent cell behaviors through specific peptide motifs that communicate with integrins and other cell membrane receptors [3]. Therefore, in order to understand how measurements of surface chemistry relate to the observed cell responses, the consideration of this layer that acts as an intermediary through which adhered cells sense the surface chemistry is of great importance [3].

The capability for a polymer surface to adsorb proteins and then facilitate cell adhesion and proliferation is related to its specific chemical composition, chemical group distribution, surface charge, surface energy, polarity, hydrophilicity, presence of certain atoms or chemical functional groups, e.g. carbon, amine groups or oxygen groups [35, 47, 55, 58, 102]. The chemical composition of the material surface is an important factor determining the surface energy, polarity, hydrophilicity and zeta potential, and consequently the character of the cell-material interaction. For example, the presence of oxygen-containing chemical functional groups increases the energy, polarity and hydrophilicity of the material surface, and supports the adhesion and growth of cells on this surface. The material surface energy, including its polar and non-polar components, can be calculated from the contact angle measurements. The hydrophilicity of the material surface can be further characterized by its zeta- (ζ-) potential [15]. Thus, these factors can be divided into three major categories: hydrophobic/hydrophilic balance, surface charge and bioactive molecule integration.

3.2.1.Hydrophobic/Hydrophilic Balance

Surface hydrophobicity has consequently been widely cited as a key factor in determining protein and cell-surface interactions [3]. On unmodified scaffolds, cell attachment occurs indirectly via a surface layer of adsorbed proteins, and the extent of protein adsorption is largely affected by surface hydrophilicity [1, 57].

It is known that hydrophobic surfaces favor the adsorption of proteins from aqueous solution thermodynamically, but may induce strongly irreversible adsorption and denature the protein's native conformation and bioactivity (a natural conformation of a protein is a prerequisite for its bioactivity). On the other hand, a highly hydrophilic surface may expel any protein molecules and inhibit protein adsorption [53].

Therefore, it is now been well accepted in biomaterial community that both very hydrophilic and hydrophobic surface will not encourage cells to attach on the surface because high surface energy results in low protein binding while low surface energy increases non-adhesive protein adsorption and denaturing of adhesive proteins [15, 35, 103-104].

It is found that surfaces with intermediate surface energy or a contact angle of ~ 50° will favor cell attachment and proliferation compared with more hydrophobic or more hydrophilic surfaces [35].

Many other works demonstrated that improved surface hydrophilicity is necessary for hydrophobic materials to support cell adherence and in particular growth, since synthetic polymers used in biotechnologies and in medicine are usually too hydrophobic in their pristine unmodified state (a water drop contact angle of about 100° or more), which cannot ensure sufficient cellular colonization [103-104].

Extensive efforts have thus been devoted towards increasing the synthetic biomaterial's hydrophilicity [1, 15, 53, 105]. Relatively safe, effective and low-cost methods for adjustment of their surface hydrophilicity are represented by physical methods, such as bombardment with ions, irradiation with ultraviolet light or exposure to plasma discharge. These procedures lead to splitting of chemical bonds between carbon and non-carbon atoms, mainly hydrogen, followed by the release of non-carbon atoms. The unsaturated carbon-carbon bonds and radicals on carbon chain react with oxygen, and the newly formed oxygen-containing groups (i.e. carbonyl, carboxyl, ester, hydroxyl groups) increase polymer surface hydrophilicity [47]. And the irradiation-activated material surface can be further functionalized by various chemical functional groups, biomolecules and nanoparticles [15]. Another widely used method is to graft hydrophilic polymers on biomaterial's surfaces through grafting copolymerization of hydrophilic polymers [53].

As mentioned earlier, the concept that moderately hydrophilic material surface induce "good" cell interaction has been widely accepted. However, this is just a very rough idea requiring further detailed studies. Few works have been contributed to detailed understanding of the optimal hydrophilicity/hydrophilicity equilibrium for a specific cell's behavior on a specific material. Recently, it is found that different optimal surface hydrophilicity is a variable value depending on cell types and specific material surfaces. For example, it was found the best water contact angle for endothelial cell attachment and proliferation is 70° on the PCL-g-PMAA [106-107], while for chondrocytes is 76° on the PLLA-g-PMAA prepared by the same grafting method [49-51].

When the water contact angle was decreased to 65°, the PLLA-g-PMAA exhibited very poor chondrocyte spreading and attachment [108]. Even for a same substrate, the grafting copolymerization methods also affect the optimal water contact angle for cell attachment.

For example, when Fe^{2+} was used to initiate the grafting polymerization after the photo-oxidization of the PLLA surface, the PLLA-g-PMAA surface showed best chondrocyte attachment and spreading when the water contact angle was 5° [109], which is much lower than 76°, the optimal value for the same PLLA-g-PMAA surface prepared by the UV induced grafting copolymerization [108]. The reason for this difference is yet to be studied. One possible reason is that the PMAA chains grafted on the PLLA surface are shorter, more uniform and of higher density when Fe^{2+} is used to initiate the grafting copolymerization, compared with the relatively longer and sparser PMAA chains when UV is used for the initiation.

The different hydrophilic PMAA chain densities and chain length may have significant influence not only on the water contact angle, but also, more importantly, on the adhesion protein molecules' conformation which will directly affect the cell attachment. In conclusion, the "moderate hydrophilicity principle" by no means gives a universally applicable optimal hydrophilicity for cell behaviors, which must differ depending on cell types and specific material surfaces [53].

On the other hand, on some hydrophilic materials, e.g. surfaces formed by plasma polymerization of acrylic acids, the absolute amount of cell adhesion-mediating ECM molecules was lower than on more hydrophobic surfaces created from octadiene [15]. Sigal*et al.* [110] studied the non-specific adsorption of several proteins to surfaces presenting different functional groups such as alkyls, amides, esters, alcohols, and nitriles. The adsorption of proteins to uncharged self-assembled monolayers (SAMs) showed a general correlation with the tendency of water to wet the surface and on the size of the proteins. While the smaller proteins (ribonuclease A and lysozyme) adsorbed only on the least wettable surfaces, larger proteins such as pyruvate kinase and fibrinogen adsorbed to some extent on almost all of the surfaces tested. There were exceptions to the general trend of increased adsorption with decreased wettability.

For example, fibrinogen adsorbed to a greater extent on SAMs presenting -CN groups than on SAMs presenting $-CH_3$ groups, although the surface presenting -CN groups has a greater wettability. Adsorption of proteins on hydrophobic surfaces was usually kinetically irreversible [56]. Lopez et al. [111] used microcontact printing to pattern gold surfaces into regions terminated in oligo(ethylene glycol) groups and methyl groups. Immersion of the patterned SAMs in solutions of the proteins such as fibronectin, fibrinogen, pyruvate kinase, streptavidin, and immunoglobulins resulted in adsorption of the proteins on the methyl-terminated regions.

Nevertheless, the cells adhered in higher numbers to more hydrophilic materials and were spread over a larger area. An explanation for these results may be that what is important is not only the absolute amount, but also the spatial conformation of the adsorbed molecules that mediate cell adhesion.

On wettable surfaces, these molecules are adsorbed in a more flexible form, which allows them to be reorganized by the cells and thus provides access for cell adhesion receptors to the adhesion motifs on these molecules. A high amount of adsorbed protein could even be disadvantageous for cell adhesion, due to denaturing of the proteins. For example, model globular proteins (hen egg lysozyme, ribonuclease A, and insulin dimer) were adsorbed to nonpolar hydrophobic surfaces, created by functionalization of self-assembled monolayers with non-polar $-CH_3$ and $-CF_3$ groups, in higher concentrations than to polar hydrophilic surfaces with PEG, -OH, or $-CONH_2$ groups.

However, the high protein concentration led to significant interprotein and protein-surface interactions. In addition, nonpolar surfaces allowed protein unfolding by reducing the unfolding free energy barriers.

As a result, a rigid adsorbed layer of proteins was formed on hydrophobic surfaces. However, on polar hydrophilic surfaces, the proteins adsorb in low concentrations, and their interprotein interactions were not significant; therefore they adsorbed in their native state.

Corresponding results were also obtained on the cell adhesion-mediating protein fibronectin. On functionalized self-assembled monolayers, the amount of adsorbed fibronectin decreased in the following order of functionalities: $NH_2 > CH_3 > COOH > OH$, while the adhesion of MC3T3-E1 osteoblast-like cells, mediated by $\alpha_5\beta_1$ integrin adhesion receptors, increased in a similar order, i.e. $CH_3 < NH_2 = COOH < OH$, which can be explained by changes in the geometrical conformation of fibronectin [15].

More investigation on the effect of surface hydrophilicity on cell adhesion and proliferation are exhibited in table 6.

Table 6. Surface hydrophilicity and the application in tissue engineering

Polymeric material	Surface chemistry	Cell adhesion and proliferation	Ref.
A hydrophilic polymer regionally fixed on a tissue culture dish and a hydrophobic polymer regionally fixed on PVA		Bovine aorta endothelial cells: adhered, spread, and confluently proliferated only on uncoated tissue culture dish surfaces or hydrophobic regions on PVA.	[110]
SAM of alkylthiols on gold with carboxylic acid and methyl termini		Human osteoblasts: after 24 h the number of cells attaching to carboxylic acid-terminated monolayers was 10 times that adherent to methyl-terminated monolayers. After 6 days in culture, osteoblasts formed multilayers on the carboxylic acid-terminated regions of the pattern.	[112]
polyethylene (PE) and PS	Ar^{+} plasma, then grafted with PEG.	Rat vascular smoothmuscle cells (VSMC): grafting with PEG increases cell adhesion and proliferation, especially on PS. Cell proliferation was shown to be an increasing function of PEG molecular weight.	[113]
PS	A series of PS surfaces created with increasing oxygen incorporation.	HUVEC: cell proliferation on surfaces increased with increasing oxygen incorporation; surface coverage with assembled fibronectin matrix was higher on the substrata with higher wettability; expression of the adhesion-related proteins cadherin-5, focal adhesion kinase and RhoA was increased on surfaces with higher wettability.	[109]
PU/PLGA films	Modified with microwave-induced argon plasma, more hydrophilic.	HUVECs: Significantly increased the cell attachment and proliferation.	[51]
Poly (D, L-lactic acid) (PDLLA)	Maleic anhydride (MAH) modified, produce pendant carboxyl groups, more hydrophilic.	Mouse osteoblast-like MC3T3-E1: extensive influence on cells by regulating expression of genes related to cell proliferation, cell cycle, cytoskeleton organization, ossification, bone remodeling, metabolism, and eventually induces osteoblast proliferation and differentiation.	[114]

3.2.2. Surface Charge

It has to note that the cell–material interactions are not only governed by the hydrophilic character of material surfaces even though moderately hydrophilic surfaces have been found to promote cell adhesion best. Other surface properties such as surface charge also have influences on cell behaviors [53, 115].

Many synthetic surfaces are charged in an aqueous environment dependent on the pKa of the surface functionalities [116]. Therefore electrostatic interactions can play an important role in determining the biological response to surfaces.

It has been shown repeatedly that there is better cell adhesion to positively charged surfaces than to negatively charged surfaces. The reason is that cell adhesion-mediating ECM molecules are negatively charged at physiological pH, thus they adsorb preferentially to positively charged surfaces [15].

Positively charged surfaces such as those containing amine groups are widely regarded to promote the adhesion of neural cells [117] and fibroblasts [118], while osteoblast adhesion on negatively charged or neutral surfaces has been shown to be limited [119]. These differences have been attributed to direct cell-surface electrostatic when protein free media is used, or the influence on protein adsorption with protein containing media [3].

Like on surfaces with different hydrophilicity, on surfaces with a different electrical charge, the spatial orientation and the biofunctionality of the adsorbed cell adhesion-promoting ECM proteins is more important than the absolute number of these molecules.

Table 7. Surface charge and the application in tissue engineering

Polymeric material	Surface chemistry	Cell adhesion and proliferation	Ref.
Poly(ethylene terephthalate) films coated with poly(benzyl N,N-diethyldithiocarbamate-co-styrene)	Three distinct ionic characters generated by potoirradiation.	Nonionic region greatly inhibited adhesion and induced clumping of the few monocytes that did not adhere. Macrophage adhesion and spreading led to high degrees of interleukin-13-induced foreign body giant cell formation on both the anionic region. In spite of the highest observed levels of monocyte/macrophage adhesion on cationic region, the adherent cells did not undergo fusion to form foreign body giant cells.	[120]
Hydrogels based on HEMA	Bearing positively charged chemical functional groups (e.g. NH^{4+}).	Stromal cells derived from rat bone marrow: adhered in significantly higher densities to positively charged than to negatively charged hydrogel surfaces.	[121]
PTFE	Irradiated with UV light in a reactive NH_3 atmosphere, the positively chargeable amine groups formed on the material surface.	Acted synergetically with oxygen-containing groups in stimulating adhesion and growth of rat aortic smooth muscle cells, mouse 3T3 fibroblasts, human umbilical vein endothelial cells and human embryonic kidney cells.	[122-124]

On self-assembled monolayers of alkanethiols terminated with positively charged $-NH_2$ group or negatively charged -COOH group, the amount of ECM protein osteopontin adsorbed to both surfaces was similar, but there is a much higher number and a much greater spreading area of bovine aortic endothelial cells on the $-NH_2$ surface than on the -COOH surface. These results suggested that the orientation and the geometrical conformation of osteopontin was more favorable for cell adhesion and spreading on a positively charged $-NH_2$ surface than on a negatively charged -COOH surface. Similarly, on surfaces patterned with positively and negatively charged microdomains, rat calvarial osteoblasts flattened better on positively charged surfaces, and mineralized tissue was also preferentially localized on positively charged regions. This cell behavior was probably mediated by a more advantageous geometrical conformation of vitronectin and fibronectin on positively charged surfaces. However, the negative charge reduced the cell-material adhesion, and also the cell-to-cell adhesion, as observed in starfish embryonic cells in situ or between endothelial cells and platelets [15].

Interestingly, negatively charged -COOH groups display a dual effect on cell-material adhesion. As relatively highly polar groups, -COOH groups can modulate the material surface wettability to values suitable for cell adhesion. In addition, -COOH groups allow grafting of various biomolecules supporting cell adhesion and growth, such as epidermal growth factor, collagen IV and chondroitin sulfate. In another study, there was also higher colonization of a copolymer of PDLLA and PEG with rat aortic smooth muscle cells on carboxylatedPEO than on non-carboxylated PEO. However, the introduction of negatively charged carboxyl groups to the adhesion substrate reduced the adhesion of human erythroleukemia cells on type I collagen matrices in vitro. Carboxyl groups on a copolymer of hydroxyethyl methacrylate and sodium methacrylate reduced the adhesion and spreading and fusion of macrophages. While, in another study, a negative charge of chemically modified films of bovine collagen did not dramatically alter the attachment of endothelial cells and fibroblasts [15].

More investigation on the effect of surface charge on cell adhesion and proliferation are shown in table 7.

3.2.3. Bioactive Molecule Integration

The effect of surface chemistry on adhering proteins and cells are largely described in terms of either the hydrophilicity or the electrostatic charge held at the surface, although more recently research in this field has highlighted the importance of specific surface functionalities, e.g. amine, carboxylic acid and hydroxyl groups. The use of more biologically relevant surfaces is also of interest, with amino acids, peptides and proteins being anchored to surfaces to control cell attachment and growth. In a signaling pathway borrowed from nature, adsorbed peptides and proteins can be used to induce cell signalling via integrin receptors atthe cell surface, first identified to control cell adhesion [3].

Based on the understanding of the dominance of the biorecognition process on cell behaviors, two main strategies in surface engineering of biomaterials are often employed. Firstly, the material surface properties such as chemical composition, hydrophilicity /hydrophobicity, surface charge and roughness, etc. are modulated to a state that the adsorbed proteins can maintain their normal bioactivities. However, this method cannot induce specific cell behaviors due to the nonspecific protein absorption. The second strategy is to directly immobilize certain biomolecules on the biomaterial surfaces to induce specific cellular responding [53]. Compared with the physical coating process previously reported, the covalent binding process via chemical reactions provides a significantly higher protein-substrate affinity, as well as a stable biological activity of the modified substrate. This will be of great interest for long-term medical devices implanted in a harsh tissue environment [1, 56-57, 102, 125].

Adding full proteins to biomaterials can be difficult to execute as the proteins are vulnerable to denaturation and degradation. A widely used method to improve cell adhesion is to incorporate short peptide motifs derived from binding regions of ECM proteins [1, 55, 102, 121, 126, 128-129]. Although they may only have a fraction of the activity of the complete protein, they can be included at very high concentration and are easy to synthesize and functionalize. Furthermore, when used in combination with PEG, non-specific protein adsorption can be eliminated and upon this blank canvas only the desired, specific receptor-mediated interactions will occur [1, 55]. As mentioned above, in more advanced biomaterials, the relatively simple, short, chemically well-defined and usually synthetic oligopeptidic or carbohydrate ligands for integrin or proteoglycan-based receptors and their cooperating sequences can be bound against a cell non-adhesive background in defined species, amount and spatial distribution in order to promote specific cell responses in a controllable manner [47]. The well-known Arg-Gly-Asp (RGD) motif from various ECM proteins including fibronectin, laminin, collagen I, fibrinogen and vitronectin was being investigated as a means of improving cell adhesion to surfaces as far back as the late 1980s [130]. It has since been incorporated into a wide range of surfaces, scaffolds and hydrogels [1, 53, 131-132]. The binding specificity of the peptide to cell surface integrin receptors can be improved by using a cyclic peptide conformation, and the flanking peptides are also critical: GRGDS is the most commonly used [133]. The inclusion of a PEG spacer arm may improve cellular response to RGD in terms of both magnitude and specificity by suppressing protein adsorption and presenting the motif to cells in a favorable fashion [134]. A biphasic cell response is noted with varying RGD concentrations: maximum cell migration is seen at intermediate ligand densities, because too few attachment sites fail to adequately support contractile forces, and high concentrations overwhelm the cells' ability to detach from the substrate [135]. Massia*et al.* determined that an average distance of 440nm between RGD peptides was sufficient to enable fibroblast attachment and spreading, but a significantly higher density (an interpeptide distance of 140nm) was required for focal contact formation and cytoskeletal reorganization into stress fibers [135]. Although RGD peptides have been extensively used, other adhesion sequences have been studied too including IKLLI, IKVAV, LRE, PDSGR and YIGSR from laminin, DGEA from collagen I and GEFYFDLRLKGDK from collagen IV, and also cell-adhesive carbohydrates [137-138]. A Pro-His-Ser-Arg-Asn (PHSRN) sequence also found in fibronectin enhances RGD activity, and the distance between the two motifs is believed to be significant for this synergistic effect. They can be separated by PEG spacers, or by glycine repeats designed to reflect the 40A° spacing between these domains in the native protein. Benoit and Anseth found that PEG-based hydrogels containing the sequence RGDG13 PHSRN improved osteoblast adhesion and spreading compared to those with just RGD [139], although elsewhere a similar sequence was shown to improve monocyte adhesion in comparison with one with 6 glycine repeats but not over RGD alone [140]. It is important to consider these precise spatial relationships when designing biomaterials. It stands to reason that cells may be expected to respond to features that exist at subcellular length scales, since their own sensing apparatus consists of molecular scale machinery [1]. It is well known that the size and shape of cell spreading area, as well as the number, size, shape and distribution of focal adhesion plaques are decisive for further migratory, proliferative and differentiation behavior of anchorage-dependent cells. If this extent is very small (i.e. attachment of round cells without formation of focal contacts and spreading), the cells usually do not survive. At the intermediate adhesion strength, the cells are most active in migration and proliferation, and if the cell-material contact area is very large with multiple well-developed focal adhesion sites associated with a rich cytoskeleton, the cells tend to skip the proliferation phase and enter sooner the differentiation program [47]. For instance, the attachment strength of fibroblasts is significantly altered by peptide spacings: spacings>90nm result in reduced focal adhesion formation, reduced detachment force and reduced cell stiffness compared to spacings<50nm suggesting that cell interaction through integrin receptors is sensitive to molecular organization on the nanoscale [141]. Another possibility how to control the proliferation and other behavior of cells on advanced biospecific materials is incorporation and controlled release of functional parts of natural growth factors, hormones, enzymes or synthetic cell cycle regulators [47, 53, 142]. The latter are represented e.g. by inhibitors of cyclin-dependent kinases (CDK), especially the compound CVT-313. This purine analog inhibits the activity of CDK2, i.e. the phosphorylation of its substrates including histone H1 or Rb protein. In someexperiments, a satisfactory restriction of cell proliferation by CVT-313 was observed in aortic smooth muscle cells derived from newborn rats, whereas the cells from adult animals were apparently less sensitive to this type of regulation [47].

4. The Bulk Modification of Polymeric Biomaterials and the Effect on Cell Adhesion and Proliferation

4.1. Composites and Their Application in Bone Tissue Engineering

Although each individual material has advantages for tissue engineering applications, each also has drawbacks associated in certain properties that can be overcome by combining different materials [20]. For instance, a key challenge in bone tissue engineering is to design a proper scaffold with the mechanical strength similar to natural bone, good biocompatibility, and adjustable biodegradability so that the scaffold could be gradually replaced with the new growing tissue. According to nanoindentation studies, human cortical bone has a Young's modulus of 5-27GPa and hardness of 0.62-0.74Gpa. These characteristics define the requirements in terms of the desired mechanical properties of bone tissue engineering scaffolds. Until now no degradable synthetic biomaterials have fully met these requirements. Incorporating of filler particles into synthetic polymeric biomaterials is a promising approach to yield a composite material with improved mechanical properties [143].

Table 8. Composite biomaterials and their application in bone tissue engineering

Composite	Preparation	Characteristic	Efficacy	Ref.
PLLA and octadecylamine-functionlizednanodiamond (ND-ODA)	Sonication dispersion, then evaporation, thin film formed.	Significantly increase in the hardness and Young's modulus, close to that of the human cortical bone.	Murine osteoblast (7F2) cell growth: no negative effects on cell proliferation.	[143]
Silicone rubber (SR) – nanohydroxyapatite (nHA)	Dispersion via shear mixing and ultrasonication, then compounding at sub-ambient temperature, and high-pressure solidification.	An improvement in modulus and tensile strength, and a retention of the high elongation of SR.	Pre-osteoblasts: initial cell attachment and proliferation consistently greater on SR-nHA than pure SR.	[145]

Table 8. (Continued).

Composite	Preparation	Characteristic	Efficacy	Ref.
PCL-HA	Wet chemical method at room temperature	NA	Bone marrow-derived human mesenchymal stem cells (hMSCs) response: supporting osteogenesis after 15days.	[146]
PLA-Ca deficient HA nanocrystals	Solution-casting technique	10.5%vol. HA had a 49% increase in the modulus.	NA	[147]
PLLA-reed fibers	Extrusion	95.6% increase in Young's modulus by adding 40%wt fibers.	NA	[148]
PCLDA-HA	Photo-crosslinking	Tensile, compressive and shear moduli greatly enhanced.	MC3T3-E1 cell attachment, proliferation and differentiation significantly enhanced.	[150]
PLLA-flax fibers	Extrusion	114.7% increase in Young's modulus by adding 40%wt fibers.	NA	[149]

The fillers studied include ceramics, natural fibers, carbon nanotubes, nanodiamond, gold nanoparticles, etc. [20, 143-144]. Different combinations of materials used to form composite biomaterials for bone tissue engineering application are presented in table 8.

4.2. Bulk Modification with Bioactive Molecules

In addition to synthesis and surface functionalization, bioactive molecules can also be incorporated into the bulk material of nanofibers/scaffolds during fabrication process. There are three most common methods of bulk modification with bioactive molecules [1, 151]. The first method is direct mixing the two solutions (bioactive molecule and polymer) before processing. It can allow for greater amounts of biomolecule incorporation and improved bioactivity when compared to surface modification techniques [152]. When two miscible solvents for a particular polymer and biomolecule solution are not available, a multi-component system may be used. For instance, a three component solution containing dichloromethance (DCM), methanol, and water was used to incorporate water soluble heparin into DCM soluble PCL nanofibers [153]. The second approach is to allow bioactive proteins to diffuse through a hydrogel from one end before crosslinking, resulting in a concentration gradient that may provide cells with directional cues [154]. Pratt et al. immobilized growth factors within hydrogels via a heparin bridge by attaching peptides with a heparin binding domain to PEG within hydrogels [155]. The third one is the insertion of bio-functional peptide sequences into peptide amphiphilic molecules for self-assembly. A peptide amphiphilicnanofiber scaffold with the incorporation of the pentapeptide epitope isolucinelysine-valine-anaine-valine (IKVAV) was designed. The effect of IKVAV incorporation in the peptide nanofibers was demonstrated by the superior bioactive properties of these scaffolds versus controls for neuronal cells [156]. Whichever of these approaches is taken, any processing steps performed after the addition of protein must allow for these sensitive molecules' need for mild conditions.

5. The Geometry of 3D Scaffolds and the Effect on Cell Adhesion and Proliferation in Bone Tissue Engineering

Bone-tissue engineering aims to heal bone defects with autologous cells and tissue without the donor site morbidity and expense associated with harvesting autogenous bone. One such strategy involves seeding autologous osteogenic cells in vitro throughout a biodegradable scaffold to create a scaffold-cell hybrid that may be called a tissue engineering construct (TEC) [157].

A key component is the scaffold that serves as a template for cell interactions and the formation of bone-extracellular matrix to provide structural support to the newly formed tissue [20]. Therefore, scaffolds for osteogenesis should mimic bone morphology, structure and function in order to optimize integration into surrounding tissue. Bone is a structure composed of hydroxyapatite ($Ca_{10}(PO_4)_6(OH)_2$) crystals deposited within an organic matrix (95% is type I collagen) [158]. The morphology is composed of trabecular bone which creates a porous environment with 50–90% porosity and pore sizes at the order of 1mm in diameter [159], with cortical bone surrounding it. Cortical bone has a solid structure with a series of voids, for example haversian canals, with a cross-sectional area of 2500–12,000μm^2 that results in 3–12% porosity [160].

There are three methods of bone regeneration using biodegradable polymers: tissue induction, cell transplantation, and fabrication of vascularized bone flaps [14]. Injectable polymers are used for filling skeletal defects and guiding bone tissue growth. Their main advantage is minimizing the surgical intervention or the severity of the surgery. Polymer-cell constructs also hold great promise in the field of tissue engineering. They provide a scaffold on which cells grow and organize themselves. As the cells begin to secrete their own extracellular matrix, the polymer degrades and is eventually eliminated from the body, resulting in completely natural tissue replacement. Bone flaps can be fabricated ectopically into precise shapes and sizes. With an attached vascular supply, these flaps can be transferred into areas deficient in vascularity.

The main challenge in preparing a useful TEC is to obtain a ubiquitous distribution of cells, and hence new tissue, throughout the entire 3D scaffold volume [157].

Traditional fabrication methods cannot build parts with predefined or controlled microstructure as well as macrostructure [161], so that existing 3D scaffolds for TE proved less than ideal for actual applications because they lack mechanical strength, interconnected channels, and controlled porosity or pores distribution [161].

Therefore, many novel processing technologies were developed, like combined extraction with phase inversion [157], rapid prototyping (RP) or solid freeform fabrication (SFF) [161], etc. The distinct advantage of these novel technologies are that they are able to build objects with predefined macrostructures as well as microstructures, exhibiting excellent potential for fabricating scaffold with controlled hierarchical structures for use in tissue engineering. The macroscopic shape of the scaffold, on a scale of up to tens of millimeters, will determine the external appearance and structure of the final product. The size, orientation, and surface chemistry of pores and channels, on a scale of hundreds of microns, have a direct impact on the extent and nature of tissue in-growth. On the scale of tens of microns, local surface texture and porosity become important [161].

5.1. Porosity and PoreSize

Porosity and pore size of biomaterial scaffolds play a critical role in bone formation in vitro and in vivo [12, 20]. Porosity is defined as the percentage of void space in a solid [163] and it is a morphological property independent of the material. Pores are necessary for bone tissue formation because they allow migration and proliferation of osteoblasts and mesenchymal cells, as well as vascularization [164]. And the necessity for porosity in bone regeneration has been shown by Kubokiet *al.* using a rat ectopic model and solid and porous particles of hydroxyapatite for BMP-2 delivery: no new bone formed on the solidparticles, while in the porous scaffolds direct osteogenesis occurred [164].

The minimum pore size required to regenerate mineralized bone is generally considered to be ~100μm after the study of Hulbert *et al.*, where calcium aluminate cylindrical pellets with 46% porosity were implanted in dog femorals [165]. Large pores (100-150 and 150-200μm) showed substantial bone ingrowth. Smaller pores (75-100μm) resulted in ingrowth of unmineralized osteoid tissue. Smaller pores (10-44 and 44-75μm) were penetrated only by fibrous tissue [165]. However, the comparison of the work from Holey *et al.* [157] and Ohgushi*et al.* [166] shows that scaffolds exhibiting similar porosity, but with a nominal pore size two orders of magnitude smaller (<10 to <100μm) could still be employed as effective hemostatic devices in which some mineralized bone-tissue islands could be identified. Moreover, Rout *et al.* [167] reported that scaffolds of even 200μm in nominal pore size would not allow cell penetration, that such pores would be overgrown by cell sheets in vitro and act as a barrier to cell and tissue invasion. Work by Akay et al. [168]shows when primary rat osteoblasts were seeded into styrene scaffolds with different pore sizes, more cells were found in the small pore (40μm) scaffolds, while cells migrated faster inside the larger pore (100μm) scaffolds; however, pore size did not affect cell penetration depth or mineralization extent. In summary, osteogenesis in vitro is not affected by pore size, but is enhanced by lower porosity. It should be noted, however, that the dimensions of pores studied are one order of magnitude lower that the minimum requirement for osteogenesis in vivo [20].

Bone regeneration in a scaffold in vivo involves recruitment and penetration of cells from the surrounding bone tissue, as well as vascularization [20]. So it is also found that the necessity for in vitro colonization by cells dictates a porosity that is quite different from that governing the in vivo response to implanted polymeric porous materials where, particularly in load-bearing conditions. Higherporosity is expected to enhance osteogenesis and numerous studies have verified this hypothesis. A porosity gradient from 80% to 88% was created in scaffolds of poly(L-lactide-co-D,L-lactide) containing 20wt% b-tricalcium phosphate (pore size 125-150μm). More tissue ingrowth and new bone formation occurred in areas with higher porosity after implantation in rabbit craniums [169]. In order to create porosity during bone healing, PPF was combined with soluble calcium filler salts; after implantation in rat tibias, scaffolds with a higher percentage of salts (and subsequently higher porosity formed in vivo) showed greater and deeper bone ingrowth [31].

In conclusion, in vitro, lower porosity stimulates osteogenesis by suppressing cell proliferation and forcing cell aggregation. In contrast, in vivo, higher porosity and pore size result in greater bone ingrowth. However, this trend results in diminished mechanical properties, thereby setting an upper functional limit for pore size and porosity. Thus, a balance must be reached depending on the repair, rate of remodeling and rate of degradation of the scaffold material. Based on early studies, the minimum requirement for pore size is considered to be ~100μm due to cell size, migration requirements and transport. However, pore sizes >300μm are recommended, due to enhanced new bone formation and the formation of capillaries. Because of vasculariziation, pore size has been shown to affect the progression of osteogenesis. Small pores favored hypoxic conditions and induced osteochondral formation before osteogenesis, while large pores, that are well-vascularized, lead to direct osteogenesis (without preceding cartilage formation). Gradients in pore sizes are recommended for future studies focused on the formation of multiple tissues and tissue interfaces [20].

In addition to porosity and pore size, the pore wall roughness and pore distribution also play a role. The effect of surface roughness has been discussed in session 3.1 before. Scaffolds with four axial and four radial channels and fixed 500μm pores separated by 500μm solid walls or 80% porous walls were fabricated from poly(desaminotyrosyl-tyrosine ethyl ester carbonate) [170]. Scaffolds from the same material with random pore distributions served as controls. Although there was no statistical difference in the bone formed in cranial defects in rabbits, bone ingrowth followed the architecture of the scaffolds: a continuous ingrowth from the outer periphery was observed in the random pore size scaffolds, while scaffolds with same sized pores and solid walls promoted discontinuous ingrowth with bone islands throughout the whole scaffold; scaffolds with same sized pores and porous walls resulted in both types of bone ingrowth. It was hypothesized that discontinuous bone ingrowth may result in faster healing, since bone will be forming not only from the margins but also throughout the whole space of the defect.

5.2. Interconnected Channels

It is known that the open pore geometry of the natural tissue is thought both to enhance metabolic activity (like nutrient transport) and to allow continuous and highly responsive bone remodeling (like cellular distribution) [171].

The morphology of some biodegradable polymer scaffolds, due to insufficient interconnected macroporosity, has been shown to limit cell colonization, and thus new tissue formation, to the superficial pore layer, resulting in tissue only ≈200μm thick [172-173]. Therefore, it is particularly important for 3D tissue engineering that tissue can form along the pore walls throughout the entire scaffold volume and not bridge across the pore interconnections, because cell layers bridging pore interconnections may prevent nutrient and fluid flow and, ultimately, tissue growth throughout the inner volume of the construct [157].

Holy, *et al.* [157] created a biodegradable scaffold from PLGA with geometry similar to that of trabecular bone by using a novel processing technique that combined extraction with phase inversion without the use of toxic solvents.

The scaffold architecture provided large pore diameters (average pore size $\approx$1.44±0.30 mm), high porosity (92%, image analysis), and large interconnections between the pores (ranged in size from 0.3 to 0.8 mm) that were separated by struts rather than by solid, membranous walls. Rodent bone-marrow cells that were seeded onto the scaffolds grew and produced mineralized bone tissue throughout the entire scaffold volume. TECs preseeded with autologous lagomorph bone marrow-derived cells were implanted into a 1.2-cm segmental femoral defect in rabbits and demonstrated new bone formation in vivo throughout the entire critical-size defect.

This marks the first report of successful 3D bone-tissue engineering repair using autologous marrow cells without the use of supplementary growth factors, indicating that the biologic strategy of a proposed therapy plays a critical role in the choice of scaffold pore size and the degree of interconnected macroporosity.

Conclusion

The field of tissue engineering places complex demands on the material it uses. The materials chosen to support the intricate processes of tissue development and maintenance need to have properties which serve both the bulk mechanical and structural requirements of the target tissue, as well as enabling interactions with cells at the molecular scale. For the construction of advanced bioartificial tissues and organs, biomaterials should not just be passively tolerated by the cells, but should actively promote specific cell responses in a controllable manner. The manner in which biomaterials interact with cells depends strongly on the physical and chemical properties of the biomaterials, and the microenvironment.

The discussion on the synthesis, surface and bulk modification of the polymeric biomaterials for cell adhesion and proliferation control, and the further application in tissue engineering has shown that cell adhesion and proliferation on synthetic polymeric biomaterials are determined by many factors, including the bulk material properties, mechanical properties, surface topography (height, depth, and shape of irregulatities; spacing between them; curvature of the valleys; sharpness of the peaks), surface chemistry (hydrophilicity, surface charge, bioactive molecular integration) and 3D scaffold geometry (pore size, porosity, interconnected channels, pore wall roughness, pore distribution). Moreover, the interaction between cell and biomaterials are also affected by cell types. Therefore, it is difficult to find an optimal value for each factors, while the possible influencing factors and the general influencing trend of each parameters introduced in the chapter could serve as a guidance for the further development of more promising polymeric biomaterials for skin, vascular, nerve and bone tissue engineering.

References

[1] E. S. Place, J. H. George, C. K. Williams and M. M. Stevens, *Chem. Soc. Rev.,*38, 1139 (2009).

[2] S. Yang, K. Leong, Z. Du and C. Chua, *Tissue Engineering*, 7 (6), 679 (2001).

[3] P. Roach, T. Parker, N. Gadegaard and M. R. Alexander, *Surface Science Reports*, 65, 145(2010).

[4] G. J. Verkerkeand E. B. v. d. Houwen, in: G. Rakhorst, R. Ploeg, Biomererials in Modern Medicine: Te Groningen Perspective, 2008, pp.23-28.

[5] E. Crubezy, P. Murail, L. Girard and J. P. Bernadou, *Nature*, 391, 29(1998).

[6] M. C. Tanzi, *Expert Review of Medical Devices*, 2, 473(2005).

[7] L L. Hench and J. M. Polak, *Science*, 295, 1014(2002).

[8] E. Nyilas, T. H. Chiu, R. L. Sidman, P. Dikkes, L. F. Borges and M. Willinger, *International Journal of Artificial Organs*, 5, 48(1982).

[9] R. Langer and J. P. Vacanti, *Science*, 260, 920(1993).

[10] J. Liu, E. E. Remsen and K. L. Wooley, *Biomacromolecules*, 2, 362(2001).

[11] E. H. Imen, M. Nakamura, M. Mie and E. Kobatake, *Journal of Biotechnology*, 139, 19(2009).

[12] M. D. Krebs, K. A. Sutter, A. S. P. Lin, R. E. Guldberg and E. Alsberg, *ActaBiomaterialia*, 5, 2847(2009).

[13] E. R. Gariepy and J. C. Leroux, *European Journal of Pharmaceutics and BIopharmaceutics*, 58, 409(2004).

[14] S. J. Peter, M. J. Miller, A. W. Yasko, M. J. Yaszemskiand A. G. Mikos, *Inc. J. Biomed. Mater. Res. (Appl. Biomater.)*, 43, 422(1998).

[15] L. Bacakova, E. Filova, M. Parizek, T. Ruml and V. Svorcik, *Biotechnology Advances*, 29, 739(2011).

[16] M. Yliperttula, B. G. Chung, A. Navaladi, A. Manbachi and A. Urtti, *European Jounal of Pharmaceutical Sciences*, 35, 151(2008).

[17] L. T. Allen, E. J. Fox, I. Blute, Z. D. Kelly, Y. Rochev, A. K. Keenan, K. A. Dawson and W. M. Gallagher, *Proc. Natl. Acad. Sci.,* 100, 6331(2003).

[18] J. Garcia, *Biomaterials*, 26, 7525(2005).

[19] H. Shin, *Biomaterials*, 28, 126(2007).

[20] V. Karageorgiou and D. Kaplan, *Biomaterilas*, 26, 5474(2005).

[21] Y. Zhang, H. Ouyang, C. T. Lim, S Ramakrishna and Z. M. Huang, *Inc. J. Biomed. Mater. Res. Part B: Appl. Biomater,* 72B, 156(2005).

[22] H. Tabesh, Gh. Amoabediny, N. S. Nik, M. Heydari, M. Yosefifard, S. O. RanaeiSiadat and K. Mottaghy, *Neurochemistry International*, 54, 73(2009).

[23] C. Zhang, N. Zhang and X. Wen, *Inc. J. Biomed. Mater. Res. Part B: Appl. Biomater,* 79B, 335(2006).

[24] S. Neuss, C. Apel. P. Buttler, B. Denecke, A. Dhanasingh, X. Ding, D. Grafahrend, A. Groger, K. Hemmrich, A. Herr, W. J. Dechent, S. Mastitskaya, A. P. Bouza, S. Rosewick, J. Salber, M. Woltje and M. Zenke, *Biomaterials*, 29, 302(2008).

[25] George and S. Ravindran, *Nano Today*, 5, 254(2010).

[26] P. Hill, H. Brantley and M. V. Dyke, *Biomaterials*, 31, 585(2010).

[27] Y. Wang, H. J. Kim, G. Vunjak-Novakovicand D. L. Kaplan, *Biomaterials*, 27, 6064(2006).

[28] P. C. Schamberger and J. A. Gardella, *Colloids and Surfaces B: Biointerfaces*, 2, 209(1994).

[29] L. Ghasemi-Mobarakeh, M. P. Prabhakaran, M. Morshed, M. H. Nasr-Esfahani and S. Ramakrishna, *Biomaterials*, 29, 4352(2008).

[30] P. J. Lou, M. Y. Chiu, C. C. Chou, B. W. Liao and T. H. Young, *Biomaterials*, 31, 1568(2010).

[31] C. Zhang, Synthesis and Characterization of Biodegradable Polyurethanes for Biomedical Application, PhD Dissertation, Clemson University, USA, (2008) .

[32] C. Zhang, N. Zhang and X. Wen, *Inc. J. Biomed. Mater. Res.*,82A, 637(2007).

[33] B. Hinz, g. Celetta. J. J. tomasek, G. gabbiani and C. Chaponnie, *Molecular Biology of the Cell*, 12, 2730(2001).

[34] M. A. Wozniak, R. Desai, P. A. Solski, C. J. Der and P. J. Keely, *Journal of Cell Biology*, 163, 583(2003).

[35] L. Cai and S. Wang, *Biomaterials*, 31, 7423(2010).

[36] Park and L. Griffith-Cima, *J. Biomed. Mate. Res.*, 31, 117(1996).

[37] J. Sagado, Y. Wang, J. F. Mano and R. L. Reis, *Mater Sci. Forum*, 514-516, 1020(2006).

[38] M. Lo, H. B. Wang, M. Dembo and Y. L. Wang, *Biophysical Journal*, 79, 144(2000).

[39] S. Nemir and J. L. West, *Ann. Biomed. Eng*., 38, 2(2010).

[40] Teixeira, S. Ilkhanizadeh, J. A. Wigenius, J. K. Duckworth, O. Inganas and O. Hermanson, *Biomaterials*, 30, 4567(2009).

[41] K. Saha, A. J. Keung, E. F. Irwin, Y. Li, L. Little and D. V. Schaffer, *Biophys. J.,* 95, 4426(2008).

[42] L. A. Flanagan, Y. E. Ju, B. Marg, M. Osterfield and P. A. Janmey, *Neuroreport*, 13, 2411(2002).

[43] P. C. Georges, W. J. Miller, D. F. Meaney, E. S. Sawyer and P. A. Janmey, *Biophys. J.,* 90, 3012(2006).

[44] Kostic, J. Sap and M. P. Sheetz, *J. Cell Sci*., 20, 3895(2007).

[45] N. D. Leipzig and M. S. Shoichet, *Biomaterials*, 30, 6867(2009).

[46] Zhang, X. Wen, N. R. Vyavahare and R. Boland, *Biomaterials*, 29, 3781(2008).

[47] L. Bacakova, E. Filova, F. Rypacek, V. Svorcik and V. Stary, *Physiol. Res*., 53(Suppl. 1), S35(2004).

[48] Shekaran and A. J. Garcia, *Biochimica et BIophysicaActa*, 1810, 350(2011).

[49] M. R. Lutolf, F. E. Weber, H. G. Schmoekel, J. C. Schense, T. Kohler, R. Muller and J. A. Hubbell, *Nat. Biotechnol*., 31, 513(2003).

[50] K. Bazaka, M. V. Jacob, R. J. Crawford and E. P. Ivanova, *ActaBiomaterialia*, 7, 2015(2011).

[51] B. J. Park, H. J. Seo, J. Kim, H. L. Kim, J. K. Kim, J. B. Choi, I. Han, S. O. Hyun, K. H. Chung and J. C. Park, *Surface and Coatings technology*, 205, 5222(2010).

[52] C. iller, A. Thapa, K. M. Haberstroh and t. J. Webster, *Biomaterials*, 25, 53(2004).

[53] Z. Ma, Z. Mao and C. Gao, *Colloids and Surfaces B. Biointerfaces*, 60, 137(2007).

[54] L. Bacakova, E. Filova, M. Parizek, T. Ruml and V. Svorcik, *Biotechnology Advances*, 29, 739(2011).

[55] Y. Ito, *Biomaterials*, 20, 2333(1999).
[56] R. S. Kane, S. Takayama, E. Ostuni, D. E. Ingber and G. M. Whitesides, *Biomaterials*, 20, 2363(1999).
[57] S. Blawas, W. M. Reichert, *Biomaterials*, 19, 595(1998).
[58] Y. M. Shin, Y. B. Lee and H. Shin, *Colloids and surfaces B: Biointerfaces*, 87, 79(2011).
[59] K. A. Thomas and S. D. Cook, *Journal of Biomedical Materials Research*, 19, 875(1985).
[60] Curtis and C. Wilkinson, *Biomaterials*, 18, 1573(1997).
[61] T. Ebendal, *Cell and tissue Research*, 175, 439(1977).
[62] P. Weiss, *Journal of Experimental Zoology*, 68, 393(1934).
[63] F. Vonrecumand T. G. Vankooten, *J BIomater. Scil. – Polym. Ed.,* 7, 181(1995).
[64] W. R. WIlkerson, C. A. Seegert, A. W. Feinberg, L. C. Zhao, M. E. Callow, R. Baneyand A. B. Brennan, *Abstr. Papers Am. Chem, Soc.*, 221, U332(2001).
[65] T. P. KUnzler, T. Drobek, M. Schuler and N. D. Spencer, *Biomaterials*, 28, 2175(2007).
[66] A.S. Crouch, D. Miller, K. L. Luebke and W. Hu, *Biomaterials*, 30, 1560(2009).
[67] X. F. Walboomers, W. Monaghan, A. S. G. Curtis and J. A. Jansen, *Journal of Biomedical Materials Research*, 14, 212(1999).
[68] M. Brunette, *Experimental Cell Research*, 164, 11(1986).
[69] M. M. Stevens and J. H. George, *Science*, 310, 1135(2005).
[70] T. Ebendal, *Experimental Cell research*, 98, 159(1976).
[71] I. Teixeira, P. F. Nealey and C. J. Murphy, *J. Biomed. Mater. Res., Part A*, 71, 369(2004).
[72] K. Yim, R. M. Reano, S. W. Pang, A. F. Yee, C. S. Chen and K. W. Leong, *Biomaterials*, 26, 5405(2005).
[73] J. Meng, Z. Han, H. Kong, X. Qi, C. Wang and S. Xie, *J. BIomed. Mater Res., A*, 95(1), 312(2010).
[74] J. Ma, X. He and E. Jabbari, *Ann. Biomed. Eng*, 39(1), 14(2011).
[75] S. H. Lim, X. Y. Liu, H. Song, K. J. Yarema and H. Q. Mao, *Biomaterials*, 31(34), 9093(2010).
[76] P. Roach, D. Farrar and C. C. Perry, *Journal of the American Chemical Society*, 128, 3939(2006).
[77] J. Y. Lim, A. D. Dreiss, Z. Zhou, J. C. Hansen, C. A. Siedleickiand R.W. Hengstebeck, *Biomaterials*, 28(10), 1787(2007).
[78] J. Y. Lim, J. C. Hansen, C. A. Siedlecki, J. Runtand H. J. Doahue, *J. R. Soc. Interface*, 2(2), 97(2005).
[79] R. L. Price, M. C. Waid, K. M. Haberstroh and T. J. Webster, *Biomaterials*, 24(11), 1877(2003).
[80] R. L. Price, K. Ellison, K. M. Haberstroh and T. J. Webster, *J. Biomed. Mater. Res., A*, 70(1), 129(2004).
[81] N. Zhang, C. Zhang and X. Wen, *Inc. J. Biomed. Mater. Res.*, 75A, 941(2005).
[82] V. T. Ribeiro-Resende, B. Koenig, S. Nichterwitz, S. Oberhoffnerand B. Schlosshauer, *Biomaterials*, 30, 5251(2009).

[83] W. M. Abbott, A. Callow, W. Moore, R. Rutherford, F. Veigh and S. Weinberg, *J. Vasc. Surg.,*17, 746(1993).

[84] Thapa, D. C. Miller, T. J. Webster and K. M. Haberstroh, *Biomaterials*, 17, 2915(2003).

[85] J. Gao, L. Niklason and R. Langer, *J. Biomed. Mater. Res.*, 3, 417(1998).

[86] Solan, V. Prabhakar and L. Niklason, *Trans. Proc.*, 33, 66(2001).

[87] R. G. Flemming, C. J. Murphy, G. A. Abrams, S. L. Goodman and P. F. Nealey, *Biomaterials,* 6, 573(1999).

[88] Thapa, D. C. Miller, T. J. Webster and K. M. Haberstroh, *Biomaterials*, 17, 2915(2003).

[89] S. amamoto, M. Tanaka, H. Sunami, E. Ito, S. Yamashita and Y. Morita, *Langmiur*, 23(15), 8114(2007).

[90] S. C. Daxini, J. W. Nichol, A. L. Sieminski, G. Smith, K. J. gooch and V. P. Shastri, *Biorheology*, 43(1), 45(2006).

[91] S. Y. Hwang, K. W. Kwon, K. J. Jang, M. C. Park, J. S. Lee and K. Y. Suh, *Anal Chem.*, 82(7), 3016(2010).

[92] E. Filova, F. Straka, L. Bacakova, M. Munzarova and J. Svobodova, Abstract No. P155, the 5th Biennial Meeting of the SHVS, Berlin, Germany, June 27-30, (2009).

[93] D. C. Miller, T. Thapa, K. M. Haberstroh and T. J. Webster, *Biomaterials*, 25(1), 53(2004).

[94] T. G. van Kooten, J. F. Whitesides and A. F. vonReçum, J. *Biomed. Mater. Res.*, 43, 1(1998).

[95] B. Cheroudi, D. McDonnel. And D. M. Brunette, *J. Biomed. Mater. Res.,* 34, 279(1997).

[96] X. F. Walboomers, H. J. E. Groes, L. A. Grinsel and J. A. Jansen, *J. Biomed. Mater. Res.,* 42, 634(1998).

[97] E. T. den Braber, J. E. de Ruijter, L. A. Ginsel, A. F. von Recum and J. A. Janse, *J.Biomed. Mater. Res*. 40, 291(1998).

[98] Z. Zhang, L. Zhang, L. Kong, A. Wang, Y. gong and X. Zhang, *J. Biomed. Mater. Res., A*. 89(2), 453(2009).

[99] K. Mtsuzaka, S. F. Walboomers, J. E. de Ruijter, J. A. Jansen, *Biomaterials*, 20, 1293(1999).

[100] Mata, . BOehm, A. J. Fleischman, G. Muschlerand S. Roy, *J. Biomed. Mater. Res.*, 62, 499(2002).

[101] D. L. Elber and J. A. Hubbell, Annu. Rev. Mater. Sci. 26, 365(1996).

[102] C. Zhang, M. E. Thompson, F. S. Markland and s. Swenson, *ActaBiomaterialia*, 7, 3746(2011).

[103] T. A. Horbett, M. B. Schway and B. D. Ratner, *J. Colloid Interface Sci*., 104, 28(1985).

[104] E. Z. Ma Tziampazis, J. Kohn and P. V. Moghe, *Biomaterials*, 21, 511(2000).

[105] Z. Ma, Z. Mao and C. Gao, *Colloids Surf.,B*, 60, 137(2007).

[106] G. Zhu, J. Ji, R. Y. Lin, C. Y. Gao, L. X. Feng and J. C. Shen, *Biomaterials*, 23, 3141(2002).

[107] Y. B. Zhu, C. Y. Gao, J. Jian and J. C. Shen, *Chin. J. Mater. Res*., 16, 233(2002).

[108] Z. W. Ma, Modification and cytocompatibility of poly-L-lactic acid (PLLA) scaffold for carlitage tissue engineering, PhD Dissertation, Zhejiang University, China, (2003).

[109] T. G. van kooten, H. T. Spijker, H. J. Busscher, *Biomaterials*, 25, 1735(2004).

[110] B. SIgal, M. Mrksich and G. M. Whitesides, *J. Am .Chem. Soc*., 120, 3464(1998).

[111] G. P. Lopez, H. A. Biebuyck, R. Hartr and A. Kumar, *J. Am. Chem. Soc.*, 115, 10774(1993).
[112] C. A. Scotchford, E. cooper, G. J. Leggett and S. Downes, J. Biomed. Mater. Res., 41, 431(1998).
[113] V. Svorcik, Z Makajova, N. Kasalkova, Z. Kolska and L. Bacakova, *J. Nanosci. Nanotechnol.* In press.
[114] Y. Xiang, Y. Wang, Y. Luo, B. Zhang, J. Xin and D. Zheng, *Colloids and Surfaces B: Biointerfaces*, 85, 248(2011).
[115] H. Lee, G. Khang, J. W. Lee and H. B. Lee, *J. Biomed. Mater. Res.,* 40, 180(1998).
[116] C. A. Haynes and W. Norde, *Colloids and Surfaces B: Biointerfaces*, 2, 517(1999).
[117] Harsch, J. Calderon, R. B. Timmons and G. W. Gross, *Journal of Neuroscience Methods*, 98, 135(2000).
[118] P. Hamerli, T. Weigel, t. Groth and D. Paul, *Biomaterials*, 24, 3989(2003).
[119] R. M. Shelton, A. C. Rasmussen and J. E. Davies, *Biomaterials*, 9, 24(1988).
[120] M. DeFife, E. Colton, Y. Nakayama, T. Matsuda and J. M. Anderson, *J. Biomed. Mater. Res.,*45, 148(1999).
[121] P. Lesny, M. Pradny, P. Jendelova, J. Michalek, J. Vacik and E. Sykova, *J. Mater. Sci. Mater. Med.*, 17(9), 829(2006).
[122] Heitz, V. Svorcik. L. Bacakova, K. Rockova, E. Ratajova and T. Gumpenberger, *J. Biomed. Mater. Res.*, 67A(1), 130(2003).
[123] R. Mikulikova, S. Moritz, T. Gumpenerger, M. Olbrich, C. Romanin and L. Bacakova, *Biomaterials*, 26(27), 5572(2005).
[124] Parizek, L. Bacakova, N. Kasalkova, V. Svorcik and K. Kolarova, *InzBiomter (Eng. BIomater.)*, X(67-68), 1(2007).
[125] H. Zisch, M. P. Lutolf, M. Ehrbar, G. P. Raeber, S. C. Rizzi, N. Davies, H. Schmokel, D. Bezuidenhout, V. Djonov, P. Zilla and J. A. Hubbell, *FASEB J.* 17, 2260(2003).
[126] Y. C. Wu, S. Y. Shaw, H. R. Lin, T. M. Lee and S. Y. Yang, *Biomaterials*, 27, 896(2006).
[127] K. P. Chennazhy and L. K. Krishnan, *Biomaterials*, 26, 5658(2005).
[128] J. W. Lee, J. H. Jung, D. S. Kim, G. Lim and D. W. Cho, *Microelectronic Engineering,* 86, 1451(2009).
[129] Z. Cheng and S. H. Teoh, *Biomaterials*, 25, 1991(2004).
[130] B. K. Brandley and R. L. Schnaar, *Anal. Biochem.* 172, 270(1988).
[131] D. H. Davis, C. S. Giannoulis, R. W. Johnson and T. A. Desai, *Biomaterials*, 23, 4019(2002).
[132] D. Klee, Z. Ademovic, A. Bosserhoff, H. Hoecher, G. Maziolis and H. J. Erli, *Biomaterials*, 24, 3663(2003).
[133] R. Haubner, R. Gratias, B. Diefenbach, S. L. Goodman, A. Jonxzykand H. Kessler, *J. Am .Chem. Soc.*, 118, 7461(1996).
[134] D. L. Hern and J. A. Hubbell, *J. Biomed. Mater. Res.*, 39, 266(1998).
[135] S. Gobin and J. L. West, *FASEB J.,* 16, 751(2002).
[136] S. P. Massia and J. A. Hubbell, *J. Cell Biol.* 114, 1089(1991).
[137] J. Salber, S. Grater, M. Harwardt, M. Hofmann, D. Klee, J. Dujic, J. H. Huang, J. D. Ding, S. Kippenberger, A. Bernd, J. Groll, J. P. Spatz and M. Moller, *Small*, 3, 1023(2007).

[138] . Heyde, M. Moens, L. van Vaeck, K. M. Shakesheff, M. C. Davies and E. H. Schacht, *Biomacromolecules*, 8, 1436(2007).
[139] D. S. W. Benoit and K. S. Anseth, *Biomaterials*, 26, 5029(2005).
[140] D. R. Schmidt and W. J. Kao, *J. Biomed. Mater. Res., Part A*, 83, 617(2007).
[141] C. Selhuber-Unkel, T. Erdmann, M. Lopez-Garcia, H. Kessler, U. S. Schwarz and J. P. Spatz, *Biophysical Journal*, 98, 543(2010).
[142] E. J. Chang, H. H. Kim, J. E. Huh, I. A. Kim, J. S. Ko, C. P. Chung and K. M. Kim, *Experimental Cell Research*, 303, 197(2005).
[143] Q. Zhang, V. N. Mochalin, I. Neitzel, I. Y. Knoke, J. Han, C. A. Klug, J. G. Zhou, P. I. Lelkes and Y. Gogotsi, *Biomaterials*, 32, 87(2011).
[144] H. S. Hung, C. C. Wu, S. Chienand S. H. Hsu, *Biomaterials*, 30, 1502(2009).
[145] W. W. Thein-Han, J. Shah and R. D. K. Misra, *ActaBiomaterialia*, 5, 2668(2009).
[146] G. Raucci, V. Danto, V. Guarino, E. Sardella, S. Zeppetelli, P. Favia and L. Ambrosio, *ActaBiomaterialia*, 6, 4090(2010).
[147] X. M. Deng, J. Y. Hao and C. S. Wang, *Biomaterials*, 22, 2867(2001).
[148] Bourmaud and S. Pimbert, *Compos. Part A Appl. Sci. Manuf.*, 39, 1444(2008)
[149] K. Oksman, M. Skrifvars, J. F. Selin, *Compos. Sci. Technol.*, 63, 1317(2003).
[150] L. Cai, A. S. Guinn and S. Wang, *ActaBiomaterialia*, 7, 2085(2011).
[151] V. Beachley and X Wen, *Progress in Polymer Science*, 35, 868(2010).
[152] H. S. Koh, T. Yong, C. K. Chan and S. Ramakrishna, *Biomaterials*, 29, 3574(2008).
[153] E. Luong-Van, L. Grondahl, K. N. Chua, K. W. Leong, V. Nurcombe and S. M. Cool, *Biomaterials*, 27, 2042(2006).
[154] C. Dodla and R. V. Bellamkonda, *J. Biomed. Mater. Res., Part A*, 78, 213(2006).
[155] B. Pratt, E. E. Weber, H. G. Schoekel, R. Muller and J. A. Hubbell, *Biotechnol. Bioeng.*, 86, 27(2004).
[156] G. A. Silva, G. Czeisler, K. L. Niece, E. Beniash, D. A. Harrington, J. A. Kessler and S. I. Stupp, *Science*, 303, 1352(2004).
[157] C. E. Holy, J. A. Fialkov, J. E. Davies and M. S. Schoichet, *Inc. J. Biomed Mater. Res.*, 65A, 447(2003).
[158] S. C. Marks, P. R. Odgren, *Principles of Bone Biology*, 2nd ed. San DiegoL Academic Press, p3-15, (2002).
[159] T. M. Keaveny, E. F. Morgan, G. L. Niebur and O. C. Yeh, *Annu. Rev. Biomed. Eng.*, 3, 307(2001).
[160] D. M. Cooper, J. R. Matyas, M. A. Katzenberg and B. Hallgrimsson, *Calcif Tissue Int.*, (2004).
[161] S. Yang, K. F. Leong, Z. Du and C. K. Chua, *Tissue engineering*, 8(1), 1(2002).
[162] M. D. Krebs, K. A. Sutter, A. S. P. Lin, R. E. Guldbergand E. Alsberg, *Acta Biomaterialia*, 5, 2847(2009).
[163] C. A. Leon, *Adv. Colloid Interface Sci.*, 76-77, 341(1998).
[164] Y. Kuboki, H. Takita, D. Kobayashi, e. Tsuruga, M. Inoue and M. Murate, *J. Biomed. Mater. Res.*, 39(2), 190(1998).
[165] S. F. Hulbert, F. A. Young, R. S. Mathews, J. J. Klawitter, C. D. Talbert, F. H. Stelling, *J. Biomed. Mater. Res.*, 4(3), 433(1970).
[166] C. T. Brighton and R. M. Hunt, *J. Bone Joint Surg. Am.* 73, 832(1991).

[167] G. J. Rout, S. F. Tarrant, J. W. Frame and J. E. Davies, *Biomaterials and Clinical Applications*, Amsterdam: Elsevier, 591(1987).

[168] G. Akay, M. A. Birch and M. A. Bokhari, *Biomaterials*, 25(18), 3991(2004).

[169] T. D. Roy, J. L. Simon, J. L. Ricci, E. D. Rekow, V. P. Thompson and J. R. Parson, *J. Biomed. Mater. Res., A*, 66(2), 283(2003).

[170] J. L. Simon, T. D. Roy, J. R. Parsons, E. D. Tekow, V. P. Thompson and J. Kemnitzer, *J. Biomed. Mater. Res., A*, 66(2), 275(2003)

[171] G. A. Rodan, *Bone*, 13, S3(1992).

[172] S. L. Ishaug, G. M. Crane, M. J. Miller, A. W. Yasko, M. J. Yaszemski and A. G. Mikos, *J. Biomed. Mater. Res,*, 36, 17(1997).

[173] S. L. Ishau-Riley, G. M. Crane-Kruger, M. J. Yaszemski and A. G. Mikos, *Biomaterials*, 19, 1405(1998).

Index

B

C

D

E

F

G

H

I

N

O

P

Q

R

S

T

U

V

W

X

Y

Z